AF784682

CONTENTS

ELECTRONIC COMMUNICATION NETWORKS IN THE WAKE OF SEPTEMBER 11

HEARING

BEFORE THE

SUBCOMMITTEE ON

COMMERCE, TRADE, AND CONSUMER PROTECTION

OF THE

COMMITTEE ON ENERGY AND COMMERCE

HOUSE OF REPRESENTATIVES

ONE HUNDRED SEVENTH CONGRESS

FIRST SESSION

DECEMBER 19, 2001

Serial No. 107–79

Printed for the use of the Committee on Energy and Commerce

Available via the World Wide Web: http://www.access.gpo.gov/congress/house

U.S. GOVERNMENT PRINTING OFFICE

77–119CC WASHINGTON : 2002

For sale by the Superintendent of Documents, U.S. Government Printing Office
Internet: bookstore.gpo.gov Phone: toll free (866) 512–1800; DC area (202) 512–1800
Fax: (202) 512–2250 Mail: Stop SSOP, Washington, DC 20402–0001

COMMITTEE ON ENERGY AND COMMERCE

W.J. "BILLY" TAUZIN, Louisiana, *Chairman*

MICHAEL BILIRAKIS, Florida
JOE BARTON, Texas
FRED UPTON, Michigan
CLIFF STEARNS, Florida
PAUL E. GILLMOR, Ohio
JAMES C. GREENWOOD, Pennsylvania
CHRISTOPHER COX, California
NATHAN DEAL, Georgia
STEVE LARGENT, Oklahoma
RICHARD BURR, North Carolina
ED WHITFIELD, Kentucky
GREG GANSKE, Iowa
CHARLIE NORWOOD, Georgia
BARBARA CUBIN, Wyoming
JOHN SHIMKUS, Illinois
HEATHER WILSON, New Mexico
JOHN B. SHADEGG, Arizona
CHARLES "CHIP" PICKERING, Mississippi
VITO FOSSELLA, New York
ROY BLUNT, Missouri
TOM DAVIS, Virginia
ED BRYANT, Tennessee
ROBERT L. EHRLICH, Jr., Maryland
STEVE BUYER, Indiana
GEORGE RADANOVICH, California
CHARLES F. BASS, New Hampshire
JOSEPH R. PITTS, Pennsylvania
MARY BONO, California
GREG WALDEN, Oregon
LEE TERRY, Nebraska

JOHN D. DINGELL, Michigan
HENRY A. WAXMAN, California
EDWARD J. MARKEY, Massachusetts
RALPH M. HALL, Texas
RICK BOUCHER, Virginia
EDOLPHUS TOWNS, New York
FRANK PALLONE, Jr., New Jersey
SHERROD BROWN, Ohio
BART GORDON, Tennessee
PETER DEUTSCH, Florida
BOBBY L. RUSH, Illinois
ANNA G. ESHOO, California
BART STUPAK, Michigan
ELIOT L. ENGEL, New York
TOM SAWYER, Ohio
ALBERT R. WYNN, Maryland
GENE GREEN, Texas
KAREN McCARTHY, Missouri
TED STRICKLAND, Ohio
DIANA DEGETTE, Colorado
THOMAS M. BARRETT, Wisconsin
BILL LUTHER, Minnesota
LOIS CAPPS, California
MICHAEL F. DOYLE, Pennsylvania
CHRISTOPHER JOHN, Louisiana
JANE HARMAN, California

DAVID V. MARVENTANO, *Staff Director*
JAMES D. BARNETTE, *General Counsel*
REID P.F. STUNTZ, *Minority Staff Director and Chief Counsel*

SUBCOMMITTEE ON COMMERCE, TRADE, AND CONSUMER PROTECTION

CLIFF STEARNS, Florida, *Chairman*

NATHAN DEAL, Georgia
Vice Chairman
ED WHITFIELD, Kentucky
BARBARA CUBIN, Wyoming
JOHN SHIMKUS, Illinois
JOHN B. SHADEGG, Arizona
ED BRYANT, Tennessee
STEVE BUYER, Indiana
GEORGE RADANOVICH, California
CHARLES F. BASS, New Hampshire
JOSEPH R. PITTS, Pennsylvania
GREG WALDEN, Oregon
LEE TERRY, Nebraska
W.J. "BILLY" TAUZIN, Louisiana
(Ex Officio)

EDOLPHUS TOWNS, New York
DIANA DEGETTE, Colorado
LOIS CAPPS, California
MICHAEL F. DOYLE, Pennsylvania
CHRISTOPHER JOHN, Louisiana
JANE HARMAN, California
HENRY A. WAXMAN, California
EDWARD J. MARKEY, Massachusetts
BART GORDON, Tennessee
PETER DEUTSCH, Florida
BOBBY L. RUSH, Illinois
ANNA G. ESHOO, California
JOHN D. DINGELL, Michigan,
(Ex Officio)

ELECTRONIC COMMUNICATION NETWORKS IN THE WAKE OF SEPTEMBER 11

WEDNESDAY, DECEMBER 19, 2001

HOUSE OF REPRESENTATIVES,
COMMITTEE ON ENERGY AND COMMERCE,
SUBCOMMITTEE ON COMMERCE, TRADE, AND CONSUMER PROTECTION,
Washington, DC.

The subcommittee met, pursuant to notice, at 10:23 a.m., in room 2322, Rayburn House Office Building, Hon. Cliff Stearns (chairman) presiding.

Members present: Representatives Stearns, Deal, Shimkus, Shadegg, Bryant, Pitts, Bass, Towns, Harman, and Markey.

Staff present: David Cavicke, majority counsel; Ramsen Betfarhad, majority counsel and policy coordinator; Shannon Vildostegui, majority counsel; Brian McCullough, majority counsel; Will Carty, legislative clerk; Jon Tripp, assistant press secretary; and Consuela Washington, minority counsel.

Mr. STEARNS. Good morning. I would like to welcome everybody to our hearing today and apologize for its lateness. We had the President come over to speak to the Republican Conference, and we were delayed and that is why we are starting a little late.

I appreciate the indulgence of the ranking member, Mr. Towns of New York, and we look forward to our witnesses.

The topic we will discuss today is a very important one. It has become more obvious since the attacks of September 11 that the decentralization in this country is good—or some would call it "sprawl" of the United States—is a huge asset in this war that has found its way to our shores. A widely dispersed population ensures viability in the unfortunate event of terror in a particular locale.

Mirroring the advantages of our population distribution, the Internet also uses an assortment of distinct connections to ensure performance even when certain locations are troubled. The Internet has a packet switch design that allows data to run on different paths until it is reassembled at its destination. Its strategic design should come as no surprise, given it was invented by the Department of Defense.

On September 11, the Internet stayed operational even with an unprecedented surge in demand. It quickly became the preferred method of communication when cellular and telephone networks became temporarily overloaded. Many Americans turned to e-mail and instant messages to communicate with loved ones and business associates.

The Web also became an important media for information-gathering. Like the television networks, news sites updated information rapidly at a time when up-to-the-minute information was extremely vital to the world.

Like the entities that provided continuity of business and personal communications on September 11, others were arguably capable of, but prevented from, performing such a mission. ECNs, as the name implies, are communication networks that facilitate commerce. At a basic level, a company like eBay is an ECN because it facilitates the meeting of buyers and sellers without the intervention of a middleman. More specialized ECNs, like our panel today will talk about, specialize in facilitating markets in stock by causing buyers and sellers to meet electronically using private electronic networks.

ECNs are electronic networks that do not have physical trading locations. Therefore, they are somewhat less susceptible to disruption of service stemming from events in a particular location. However, a catastrophic incident of the magnitude of September 11 can still affect the communication infrastructure that the ECNs depend on.

The technology of the ECNs is a development that reflects the changing nature of all commerce, both domestic and, of course, global. An ECN is not unlike the Internet in that it provides a platform that allows perfect strangers to enter from anywhere and meet in an anonymous environment. In this particular case, they meet in order to trade stock. Although they are quite different from the traditional exchanges, the ECNs can provide a unique alternative to the markets. However, securities regulations prevent these entities from operating when the traditional exchanges are closed.

Following the attacks of September 11, the markets were closed for 4 business days and no rules—excuse me—no trades were conducted. I understand the decision was based in part on both security concerns and the loss of power for market participants and their ability to connect to these markets. Additionally, even though it appears some of these ECNs had the ability to operate in a situation such as the market shutdown, the markets remained closed. It would have been impractical to trade when the consolidated tape was closed for 4 days.

My colleagues—on a related note, the committee has had correspondence last year with the SEC on an issue that affects ECNs, the rules governing the Consolidated Tape Association, or CTA. The CTA, as I understand it, is a group that splits the fees from market data generated by stock trades. This data is very valuable to ordinary people who wish to buy or sell stocks. To change the rules of CTA or to admit new members requires the unanimous consent of all the current members. Because of our dealings with the Senate, we know that this unanimous consent is difficult to achieve. That is why we are here today.

I would like our witnesses to address what changes they think could be made to the CTA to make it a more modern and efficient organization. So I look forward to learning more about these issues and considering the suggestions from our witnesses on shaping a stronger market that can utilize the advantages of today's technology.

And, with that, the opening statement from our distinguished ranking member, Mr. Towns of New York.

[The prepared statement of Hon. Cliff Stearns follows:]

PREPARED STATEMENT OF HON. CLIFFORD STEARNS, CHAIRMAN, SUBCOMMITTEE ON COMMERCE, TRADE, AND CONSUMER PROTECTION

I would like to thank our panel of witnesses for being here today. The topic we will discuss is an important one. It has become more obvious since the attacks that the decentralized nature, or as some call "sprawl" of the United States is a huge asset in this war that has found its way to our shores. A widely dispersed population ensures viability in the unfortunate event of terror in a particular locale. Mirroring the advantages of our population distribution, the Internet also uses an assortment of distinct connections to ensure performance even when certain locations are troubled. The Internet has a packet switch design that allows data to run on different paths until it is reassembled at its destination. Its strategic design should come as no surprise given it was invented by the Department of Defense.

On September 11th, the Internet stayed operational even with an unprecedented surge in demand. It quickly became the preferred method of communication when cellular and telephone networks became temporarily overloaded. Many Americans turned to email and instant messages to communicate with loved ones and business associates.

The Web also became an important medium for information gathering. Like the television networks, news sites updated information rapidly at a time when up to the minute information was vital to the world.

Like the entities that provided continuity of business and personal communications on the 11th, others were arguably capable of, but prevented from, performing such a mission. Electronic Communications Networks, or ECNs, are entities that provide an electronic platform for trading securities. As their name implies, ECNs are electronic networks that do not have physical trading locations. Therefore, they are somewhat less susceptible to disruption of service stemming from events in a particular location. However, a catastrophe the magnitude of September 11 can still affect the communication infrastructure that the ECNs depend on.

The technology of the ECNs is a development that reflects the changing nature of all commerce—both domestic and global. An ECN is not unlike the Internet in that it provides a platform that allows perfect strangers to enter from anywhere and meet in an anonymous environment. In this particular case, they meet in order to trade stock. Although they are quite different from the traditional exchanges, ECNs can provide a unique alternative to the markets. However, securities regulations prevent these entities from operating when the traditional exchanges are closed. Following the attacks of September 11th, the markets were closed for four business days and no trades were conducted. I understand the decision was based in part on both security concerns and the loss of power for market participants and their ability to connect to markets. Additionally, even though it appears some of the ECNs had the ability to operate in a situation such as the market shutdown, the markets remained closed. It would have been impracticable to trade when the consolidated tape was closed for four days.

On a related note, I would like our witnesses to address another issue with the consolidated tape. It is my understanding that the Consolidated Tape Association requires unanimity for rule changes to the operations of the system. I find this troubling as one participant can hold up advances in the system. I would be grateful if the witnesses would address solutions to this problem.

I look forward to learning more about these issues and considering our witnesses' suggestions on shaping a stronger market that can utilize the advantages of today's technology.

Mr. TOWNS. Thank you very much, Mr. Chairman, and also thank you for holding this hearing.

The Nasdaq stock market and the New York Stock Exchange should be commended for the heroic efforts that their staffs performed to allow for the opening of their markets within 6 days after the attacks. The unified action taken by them, together with the American Stock Exchange and the SEC and New York State and city officials, should serve as a model for future public-private undertakings.

It is truly a shame that it takes events like September 11 to bring the best out of us. I am hopeful that this new spirit can be a permanent part of our financial landscape.

While this hearing is focused on ECNs, it would be a mistake to overlook or minimize the value of the primary markets and all they contribute to capital formation and economic improvement in this country. They are truly natural resources and enjoy the envy of the rest of the world. Moreover, this committee has jurisdiction over the computer systems and telecommunications infrastructure that forms the backbone of ECNs and exchanges, indeed, the entirety of the securities industry.

I firmly believe that you all provide important services to investors. I deeply regret the losses that you sustained, be they personal or physical, in the September 11 attacks. I look forward to hearing from you about lessons that we can learn and what steps we should be taking, going forward, to improve our response to such disasters. I am especially interested in your suggestions regarding systems security and continuity planning.

Mr. Chairman, on that note, I yield back.

Mr. STEARNS. I thank my colleague.

The gentleman from New Hampshire, Mr. Bass.

Mr. BASS. Thank you very much, Mr. Chairman. I thank you for holding this fascinating hearing. I really didn't understand this issue until a week or so ago, and I think it is yet another example of how modern technology is moving forward to dramatically expand opportunities for commerce in this country. And as you mentioned in your opening statement, what platforms like eBay and others do for the auction business, so electronic communications networks may do for all sorts of other commerce-related issues, especially securities trade and so forth.

As you also mentioned in your opening statement, there are, however, issues that may need to be addressed in order to protect consumers and to make sure that commerce that does occur through this medium is done in an orderly and reliable and secure manner.

So I welcome the witnesses that we have here today. This is going to be a learning experience for me, and I appreciate your holding this hearing. I yield back.

Mr. STEARNS. I thank the gentleman.

Mr. BASS. Can I submit a statement for the record?

Mr. STEARNS. By unanimous consent, so ordered.

[The prepared statement of Hon. Charles F. Bass follows:]

PREPARED STATEMENT OF HON. CHARLES BASS, A REPRESENTATIVE IN CONGRESS FROM THE STATE OF NEW HAMPSHIRE

Mr. Chairman, thank you for holding this hearing. As others have pointed out, September 11 gave our system of electronic communications quite a test.

Most of us were part of that test as we rushed to our landline and mobile phones, our pagers, our PDAs, and our email accounts. Although phone service from our offices was intermittently down or overloaded, and the mobile networks were incapable of handling the volume, Internet based systems performed well due to their packet switching technology.

I don't know what the proper role is for Congress to play in ensuring continuity of service during a crisis. Certainly, several items stand out as needing review. For example, why weren't emergency broadcasting systems used to better effect? Why were those of us in this area unable to have the numerous false rumors disproven for so long? And why weren't we able to more effectively ask the public to use the communications network sparingly, thus keeping it free for emergency use? I hope that this hearing will, in part, begin to answer these and other questions.

On September 24, I visited southern Manhattan with several members of this committee. We saw for ourselves the extent of the horror and damage to the area. While there, we toured the site of the World Trade Center and the nearby Verizon building at 140 West Street. Like so many other nearby facilities, it was impossible to imagine that the building would be able to perform any serviceable action whatsoever.

Nevertheless, Verizon and many other service providers undertook absolutely monumental efforts to get the system running as quickly and confidently as possible. Each of these communications workers deserves our thanks—and they certainly have mine.

I look forward to hearing from these witnesses, and I yield back to the Chairman.

Mr. STEARNS. The gentleman from Tennessee, Mr. Bryant.

Mr. BRYANT. Thank you, Mr. Chairman. I think Mr. Pitts actually arrived before I did, but I will be brief.

I want to welcome this panel. And like Mr. Bass, I think he has probably already learned enough about this, but there is more I need to learn. But I think neither of us knew very much before we started. I welcome you.

And I thank you for holding this hearing and I would yield back the balance of my time.

Mr. STEARNS. Mr. Pitts?

Mr. PITTS. I have no opening statement.

Mr. STEARNS. Mr. Shimkus.

Mr. SHIMKUS. Just a brief follow-up. I know Chairman Upton in the telecom subcommittee visited Ground Zero about 4 days after. One of the reasons why was to observe the enormous amount of work that went—and I don't think the public really understands what was accomplished to get the financial markets back on line. It is a tremendous, tremendous success story; but as much as a success story, it is also a warning on how do we have the infrastructure to make sure that we can respond and continue to operate in a manner that provides confidence.

I think that was one of the—of all the tremendous things that occurred in the visit and seeing everything, the fact that the economic stability of the country was maintained with such a devastating blow to a large portion of the financial sector, and the communication aspects and the ability to get up and running and really without—I don't think the vast majority of the public understands and saw no difference after the delay and when the markets reopened. It is an incredible statement.

But we do have to look at how we can make sure, how we are able to do that in the future and what type of systems need to be in place.

Mr. Chairman, I thank you for the hearing. I think it is going to be educational, eye-opening; and hopefully, even for the constituents across the country, they will understand the incredible amount of work that went on to help provide economic stability for this country in a tremendous time of crisis.

And with that, I yield back the balance of my time.

Mr. STEARNS. I thank the gentleman.

With no further opening statements, we will move to our panel, which again, I want to welcome all of you. We have Mr. Joe Steinmetz, senior vice president of—yes, Mr. Towns?

Mr. TOWNS. I would like unanimous consent that other members who are not here also have an opportunity to put their statements in the record.

Mr. STEARNS. So ordered.

[Additional statements submitted for the record follow:]

PREPARED STATEMENT OF HON. JOHN SHADEGG, A REPRESENTATIVE IN CONGRESS FROM THE STATE OF ARIZONA

Thank you Mr. Chairman for holding this important hearing today. Let me first express my deepest sympathy to those in the financial services sector who lost employees in the terrible attack on the World Trade Center towers. Words cannot describe the tragedy and loss.

I also want to applaud the tremendous effort of all who were involved in putting the telecommunications infrastructure back in place in order to get trading back up in such a short time after the 9/11 tragedy.

Mr. Chairman, today, half of all Americans own stocks, up from 36 percent in the early 1990s. At the same time, however, I would guess that a very small fraction of that percentage understand Electronic Communications Networks and how they may change the future of investing.

So, it is with anticipation Mr. Chairman, that I look forward to learning more today about the telecommunications infrastructure that supports ECNs and the regulatory structure that may or may not need to be changed to keep pace with technological innovation.

I yield back.

PREPARED STATEMENT OF HON. W.J. "BILLY" TAUZIN, CHAIRMAN, COMMITTEE ON ENERGY AND COMMERCE

Chairman Stearns, thank you for holding this hearing this morning, a hearing that will help us understand how we might improve a vital aspect of electronic commerce.

As the name of this Committee implies, commerce is the cornerstone of our jurisdiction. Since the September 11 tragedies, we have been spending quite a bit of time determining how various industries essential to commerce fared in the wake of the attacks. We have spent time examining how these industries are prepared to deal with future shocks to the system. As the familiar old cliché perfectly states: "Failing to prepare is preparing to fail."

There is ample evidence that, in the aftermath of the attacks, most industries' inability to operate as normal was only temporary. However, temporary interruptions still had severe results. We learned last month of losses suffered by the travel and tourism industry from the drop-off in air travel. Unfortunately, it has been a hard road back for that industry.

The securities markets, too, were among the most severely affected by the attacks. The policy decision was made to close the markets for several days. While circumstances were extraordinary, it is disconcerting that the markets in this day and age were shut down as long as they were.

As our economy continues to evolve into an electronic marketplace, the fundamental principle of commerce that we must protect is the ability to exchange information as efficiently and reliably as possible. Continuity of operations is part of this equation.

Fortunately, with the technology and communication advancements of the past decade, electronic communications networks—ECNs—have filled that vital role and emerged as viable conduits to the securities markets. Many have made a niche in continuing trading in the hours after markets are traditionally closed. By meeting the demands of investors in creative and efficient ways, ECNs have sparked competition and innovation that has improved our marketplace and benefited consumers and the businesses that rely on access to capital.

The purpose of the hearing today is to identify any barriers that may prevent the technology at our witnesses' disposal from being used more broadly to benefit of investors. As the world leader of free markets, the United States must make sure that regulation serves to make technology an asset to strong markets—not stand as an impediment.

I thank the participants for coming today to share their views with the Committee and look forward to continuing to work with you as we discuss these matters of public policy.

Mr. STEARNS. [continuing] Instinet of New York City.

We have Mr. Matthew Andresen, President and CEO of The Island ECN. We have Ms. Catherine Kinney, Group Executive Vice President of the New York Stock Exchange.

And I understand you are going to be the President. Congratulations.

Mr. Kim Bang, President of Bloomberg Tradebook, New York City; Mr. Kevin O'Hara, General Counsel, Archipelago, LLC; Mr. Steven Randich, Executive Vice President of Operations and Technology, Chief Information Officer of the Nasdaq; and last, Mr. Keith Jamaitis, Senior Vice President, Chief Operating Officer of NYFIX.

So, with that, we are very pleased to have your opening statement. And we will start with you, Mr. Steinmetz.

STATEMENTS OF JOEL STEINMETZ, SENIOR VICE PRESIDENT, INSTINET; MATTHEW ANDRESEN, PRESIDENT AND CHIEF EXECUTIVE OFFICER, THE ISLAND ECN; CATHERINE R. KINNEY, GROUP EXECUTIVE VICE PRESIDENT, NEW YORK STOCK EXCHANGE, INC.; KIM BANG, PRESIDENT, BLOOMBERG TRADEBOOK; KEVIN J.P. O'HARA, GENERAL COUNSEL, ARCHIPELAGO, LLC; STEVEN J. RANDICH, EXECUTIVE VICE PRESIDENT OF OPERATIONS AND TECHNOLOGY, CHIEF INFORMATION OFFICER, THE NASDAQ STOCK MARKET, INC.; AND KEITH R. JAMAITIS, SENIOR VICE PRESIDENT, NYFIX MILLENIUM

Mr. STEINMETZ. Thank you.

Mr. Chairman and members of the subcommittee, thank you for the opportunity to appear before you this morning to discuss the impact of the events of September 11 upon electronic communications networks. My name is Joel Steinmetz and I am Senior Vice President, responsible for equities at Instinet Corporation.

Instinet is the world's largest and oldest electronic agency securities broker and has been providing investors with electronic trading solutions for more than 30 years. Instinet's clients—mutual funds, pension funds, insurance companies, corporations and market professionals—represent more than 90 percent of U.S. Managed institutional stock funds. Instinet is a member of 20 exchanges in North America Europe and Asia, and our clients use our services to trade on more than 40 markets around the world. Last year, Instinet's customers used its systems to execute almost 88 million transactions globally of which almost 83 million transactions were in U.S. Equity securities.

As an important part of its services to its clients, Instinet acts as an electronic communications network, or ECN. ECNs are electronic marketplaces that allow institutional, retail and professional participants to trade securities directly with one another, as well as with other securities firms. ECNs are operated by companies that are registered with the Securities and Exchange Commission as broker-dealers and that are members of the National Association of Securities Dealers.

ECNs provide electronic agency brokerage services, meaning that they match customer orders as agents, not principals. In other words, an order is executed if a matching order is immediately available from another customer of the ECN. If no matching order is available, the order is displayed in the electronic order book and becomes eligible for execution by orders entered by other subscribers. ECNs typically are compensated by small commissions paid equally by the seller and buyer in each transaction, generally on a per-share basis.

Another consequence of being an agency broker is that, unlike Nasdaq market-makers or exchange specialists, ECNs do not trade for their own accounts. In recent years, the SEC has established special additional regulatory requirements applicable to ECNs as well as to other alternative trading systems.

ECNs function both through the operation of proprietary internal networks that connect their customers to the ECNs own systems and through external networks that connect them to other ECNs and other market participants. The proprietary systems are used to display the available order book, offers of sales or purchases at stated prices and volumes, as well as to provide connectivity to other systems that execute the trades once they match. ECNs may also offer other products and services through their networks, including research and analysis.

ECNs provide important services to issuers and investors in the financial marketplace. First, ECNs allow their customers to trade with one another directly. Second, ECNs do not trade their own accounts. They are completely neutral with regard to their clients trading strategies. Together these factors allow ECNs to create efficiencies that can significantly improve their client's trading prices and reduce overall transaction costs.

In addition, ECNs do not require the identity of the ultimate buyer or seller to be disclosed to the other market participant at any point in the trading process. This anonymity can reduce the potential market impact of large transactions and transactions by certain investors whose trading activity, if known, may be more likely to influence other market activity.

Another benefit that ECNs provide is direct access to markets for all of their customers, which can increase the speed at which trades are executed and can level the playing field among market participants.

In short, ECNs have been leaders of innovation in bringing technological advances and using those advances for the benefit of consumers. Among the advances we are now developing, as I will discuss shortly, is increased use of distributed systems, which allows for more robust trading systems and could help in the future to absorb shocks.

Turning to the events of September 11, I would first like to take the opportunity on behalf of Instinet to extend my deepest sympathy to all who suffered and continue to suffer in the wake of that unspeakable tragedy. I would also like to express any admiration and thanks to the countless people whose courage, strength and determination have helped our country get through this difficult time.

Instinet, too, was directly affected by the events of September 11. Tragically, we lost two Instinet colleagues who were attending a conference in the World Trade Center at the time of the attacks. Everyone in the Instinet community has felt this loss deeply.

Fortunately, our other employees who worked at the World Trade Center were able to escape. Instinet had important facilities located in the World Trade Center that were destroyed by the attacks, thus Instinet's operations were disrupted but not solely due to the loss of our own facilities. Specifically, 1 of our 3 principal data centers, through which many of our customer communications were routed, were destroyed in the collapse. As a result, we lost connectivity with those customers whose sole access to our system was through the data center in the World Trade Center. Our clearing operations were also housed in the World Trade Center and were also destroyed.

In addition, however, we were affected by the destruction of Verizon's West Street central office across from the Towers. In addition to the direct impact of the loss of Verizon's service to us, the Verizon outage also caused us to be disconnected from some of our customers on Verizon's network who otherwise connected, or could connect, to one of our two other principal data centers. These losses in combination resulted in our losing connectivity to about one-third of our customer base immediately after the attacks.

Fortunately, however, due to the hard work of Verizon, government regulators and, above all, our employees, we were able to restore connectivity with most of our customers in remarkably short time. We did this through a variety of means.

For example, in the days after the attacks, we worked to circumvent the damaged units and by utilizing alternative lines. Customers that previously connected to the World Trade Center data center were rerouted to or through our centers in New Jersey and Boston, and we made use of ISDN lines in some instances. Also, we accelerated a pilot project we already had in place to provide access through the Internet, which provided an alternative and quite effective means to restore connections for some customers where it would have taken longer to restore standard methods of communication.

By using these alternative communication routes, we were able to restore connectivity to approximately 90 percent of our customers by the end of the week.

It is important to note, however, that our ability to conduct trading itself was not disrupted. Our customers in Europe and Asia, for example, were able to resume trading the day after the attacks; and customers in the United States that had not lost connectivity to us were able to view price information on the system that day as well. The principal constraint on trading in the United States was the need for settlement services and for the interconnection with other market segments, in addition to restoration of connectivity to all of our customers.

In the weeks that followed, we began to focus not only on rebuilding our damaged infrastructure, but also on improving on it going forward. For instance, we sought to identify and eliminate potential single points of failure that could, in the event they were impaired, result in a loss of connectivity or other functionality. Our efforts in this regard have proceeded rapidly and are already about 70 percent of the way there.

In addition, we are creating a more resilient and robust network, one which we expect will be less subject to disruption in the event of future catastrophic events. We are achieving this by, among other things, increasing the number of available connections with our customers.

The events of September 11 and their impact upon our business have taught us a number of important lessons. Even before September 11, we placed a high premium on redundant systems to permit continuous service at all times. The unprecedented damage caused on September 11 underscored the necessity of backup systems, redundant systems and contingency planning.

Mr. STEARNS. We will need you to sum up, if you could.

Mr. STEINMETZ. Such systems and backup plans must be thorough and reliable and capable of being implemented on short notice. In general, less centralized systems like the Internet are better able to absorb and respond to disruption and devastating events than are more centralized ones. The less centralized systems thereby minimize interruption to the many affected parties.

Indeed, the attacks did not impair Internet communications at all. Had Instinet's system of Internet access already been fully operational, there likely would have been significantly less disruption to our customer connectivity. Moreover, a network-style market structure provides a better framework to encourage competition and innovation through the use of emerging technologies.

This has been Congress' goal since the 1975 amendments to the Securities Exchange Act of 1934; and it is one Instinet strongly promotes. Thank you.

[The prepared statement of Joel Steinmetz follows:]

PREPARED STATEMENT OF JOEL STEINMETZ, SENIOR VICE PRESIDENT, EQUITIES, INSTINET CORPORATION

I. INSTINET AND ECNS

Mr. Chairman and members of the Subcommittee, thank you for the opportunity to appear before you this morning to discuss the impact of the events of September 11 upon electronic communications networks. My name is Joel Steinmetz and I am Senior Vice President for Equities at Instinet Corporation.

Instinet is the world's largest and oldest electronic agency securities broker, and has been providing investors with electronic trading solutions for more than 30 years. Instinet's clients—mutual funds, pension funds, insurance companies, corporations, and market professionals—represent more than 90% of U.S.-managed institutional stock funds. Instinet is a member of 20 exchanges in North America, Europe, and Asia and our clients use our services to trade in more than 40 markets around the world. Last year, Instinet's customers used its systems to execute almost 87.6 million transactions globally, of which 82.4 million transactions were in U.S. equity securities.

As an important part of its services to clients, Instinet acts as an electronic communications network, or "ECN." ECNs are electronic marketplaces that allow institutional, retail and professional market participants to trade securities directly with one another, as well as with other securities firms. ECNs are operated by companies that are registered with the Securities and Exchange Commission as "broker-dealers" and that are members of the National Association of Securities Dealers. ECNs provide electronic agency brokerage services, meaning that they match customer orders as agents, not principals. In other words, an order is executed if a matching order is immediately available from another customer on the ECN. If no matching order is available, the order is displayed in the electronic order "book" and becomes eligible for execution by orders entered by other subscribers. ECNs typically are compensated by commissions paid by the seller and buyer in each transaction, generally on a per share basis. Another consequence of being an agency broker is that, unlike Nasdaq market makers or exchange specialists, ECNs do not trade for their own accounts. In recent years, the SEC has established special additional regulatory requirements applicable to ECNs, as well as to other "alternative trading systems."[1]

ECNs function both through the operation of proprietary internal networks that connect their customers to the ECN's own systems and through external networks that connect them to other ECNs and other market participants. The proprietary systems are used to display the available order "book"—offers of sales or purchases at stated prices and volumes that have not yet been matched—as well as to provide connectivity to other systems that execute the trades once they match.[2] Instinet also offers other products and services through its network, including research, analytics, and "smart routing" of customers' orders to other execution venues.

ECNs provide important services to issuers and investors in the financial marketplace. First, ECNs allow their customers to trade with one another directly. Second, ECNs do not trade for their own accounts—they are completely neutral with regard to their clients' trading strategies. Together these factors allow ECNs to create efficiencies that can significantly improve their clients' trading prices and reduce overall transaction costs. In addition, ECNs do not require the identity of the ultimate buyer or seller to be disclosed to the other market participants at any point in the trading process. This anonymity can reduce the potential market impact of large transactions and transactions by certain investors whose trading activity, if known, may be more likely to influence other market activity. Another benefit that ECNs provide is direct access to markets for all of their customers, which can increase the speed at which trades are executed and can level the playing field among market participants.

In short, ECNs have been leaders of innovation in bringing technological advances to securities trading and using those advances for the benefit of consumers. Among the advances we are now developing, as I will discuss shortly, is increased use of decentralized systems, which allows for more robust trading systems and could help in the future to absorb shocks.

II. THE IMPACT OF THE EVENTS OF SEPTEMBER 11 UPON INSTINET

Turning to the events of September 11, I would first like to take the opportunity, on behalf of Instinet, to extend my deepest sympathy to all who suffered, and continue to suffer, in the wake of that unspeakable tragedy. I would also like to express my admiration and thanks to the countless people whose courage, strength, and determination have helped our country get through this difficult time.

Instinet, too, was directly affected by the events of September 11. Tragically, we lost two Instinet colleagues who were attending a conference in the World Trade Center at the time of the attacks. Everyone in the Instinet community has felt this loss deeply. Fortunately, our other employees who worked at the World Trade Center were able to escape.

Instinet had important facilities located in the World Trade Center that were destroyed by the attacks. Thus, Instinet's operations were disrupted, but not solely due to the loss of our own facilities. Specifically, one of our three principal data centers, through which many of our customer communications were routed, was destroyed in the collapse. As a result, we lost connectivity with those customers whose sole access to our system was through the data center in the World Trade Center. Our clearing operations were also housed in the World Trade Center and were also destroyed. In addition, however, we were affected by the destruction of Verizon's West Street central office across from the towers. In addition to the direct impact of the loss of Verizon's service to us, the Verizon outage also caused us to be disconnected from some of our customers on Verizon's network who otherwise connected, or could connect, to one of our two other principal data centers. These losses in combination resulted in our losing connectivity to about one-third of our customer base immediately after the attacks.

Fortunately, however, due to the hard work of Verizon, government regulators, and, above all, our employees, we were able to restore connectivity with most of our customers in a remarkably short time. We did this through a variety of means. For example, in the days after the attacks, we worked to circumvent the damaged units and to utilize alternative lines. Customers that previously connected to the World Trade Center data center were re-routed to or through our centers in New Jersey and Boston, and we made use of ISDN lines in some instances. Also, we accelerated a pilot project we already had in place to provide access through the Internet. This

[1] These include in particular the Order Handling Rules and Regulation ATS.

[2] Pursuant to the SEC's Order Handling Rules, ECNs display their customers' best-priced buy and sell orders to Nasdaq and provide access to those orders through Nasdaq to NASD members who are not ECN subscribers. Non-subscribers accessing ECN orders may be charged a fee, but it is capped at 1.5 cents per share.

provided an alternative and quite effective means to restore connections for some customers where it would have taken longer to restore our standard methods for connection.

By using these alternative communication routes, we were able to restore connectivity to approximately 90% of our high-volume customers by the end of September 13. Within a day of the equity markets reopening on September 17, we had restored connectivity to 75% of all our customers and were at the 95% level two days later. It is important to note, however, that our ability to conduct trading itself was not disrupted. Our customers in Europe and Asia, for example, were able to resume trading in non-U.S. securities the day after the attacks, and customers in the United States that had not lost connectivity to us were able to view price information on the system that day as well. The principal constraint on trading in the United States was the need for settlement services and for interconnection with other market segments, in addition to restoration of connectivity to all our customers.

In the weeks that followed, we began to focus not only on rebuilding our damaged infrastructure, but also on improving it going forward. For instance, we sought to identify and eliminate potential single points of failure that could, in the event they were impaired, result in a loss of connectivity or other functionality. Our efforts in this regard have proceeded rapidly and are already about 70% of the way there in implementing these improvements. In addition, we are creating a more resilient and robust network, one which we expect will be less subject to disruption in the event of future catastrophic events. We are achieving this by, among other things, increasing the number of "backup" or "redundant" connections with each of our customers.

III. LESSONS LEARNED IN THE WAKE OF SEPTEMBER 11

The events of September 11 and their impact upon our business have taught us a number of important lessons. Even before September 11, we placed a high premium on redundant systems to permit continuous service at all times. The unprecedented damage caused on September 11 underscored the necessity of backup systems, redundant systems, and contingency planning. Such systems and backup plans must be thorough, reliable, and capable of being implemented upon short notice.

The disruption caused by the attacks also demonstrated the value of "network" or decentralized systems, such as the Internet, relative to older, more centralized "mainframe"-type communications and trading systems. Indeed, the attacks did not impair Internet communications at all. Similarly, more decentralized financial markets like the currency markets and bond markets suffered disruption equal to or greater than the stock markets, but were able to resume trading more quickly. In general, less centralized systems like the Internet are better able to absorb and respond to disruption and devastating events than are more centralized ones. The less centralized systems thereby minimize interruption to the many affected parties. Moreover, a network-style structure provides a better framework to encourage competition and innovation through the use of emerging technologies.

Instinet believes that "network"-style communications systems hold continuing promise for the equities markets as well. Moving from the current system, with its mandatory linkages and centralized information monopoly, to a network-style structure would produce benefits for investors. Competition, spurred by advances in information technology, has substantially improved the transparency of the securities markets over the last several decades. Investors have a variety of options for the display and execution of their orders. Competition has also led market participants to develop voluntary linkages among themselves. These facilitate execution of investors' orders in other market centers while preserving the freedom of individual systems to innovate. A decentralized, "open architecture" structure would provide the greatest scope for competition and innovation to operate, benefiting investors through lower costs, better services, and more efficient trading. These have been Congress's goals since the 1975 amendments to the Securities Exchange Act of 1934, and they are goals Instinet strongly promotes.

Thank you.

Mr. STEARNS. Thank you.

Mr. Andresen.

STATEMENT OF MATTHEW ANDRESEN

Mr. ANDRESEN. Thank you, Chairman Stearns, Ranking Member Towns and the members of the subcommittee.

I would like first to commend the chairman and the members of the subcommittee for holding these hearings concerning the tragic events of September 11 and specifically the role of electronic communications networks in our securities markets.

I am Matthew Andresen. I am the Chief Executive Officer of Island ECN. Island is an electronic marketplace that enables market participants to display and match orders for stocks and other securities. Island's proprietary technology allows it to offer this low-cost, rapid and reliable order display and matching service to a network of approximately 700 broker-dealers representing a diverse array of market participants.

On an average day, Island will trade over 460 million shares and over 1 in every 5 trades on Nasdaq. Through November of this year, Island has traded over 82 billion shares worth $2.5 trillion. Indeed, recent analyst reports indicate that Island is now the largest ECN in the world today.

Mr. Chairman, unquestionably the events of September 11 have had a profound impact on our whole Nation, not just on the securities markets. Those in the New York financial community, however, have felt an additional obligation to prove to the world the strength and the resilience of America's financial markets.

I am proud that despite the initial shock of the attacks and even the fear of further attacks, Island has rebounded quickly. From the employees who were on the roof that morning cleaning out air conditioning ducts while burning soot blackened the whole of Wall Street to a team of technical staff that literally worked around the clock to reestablish connectivity to our subscribers whose lines ran through Ground Zero, the Island staff rose to the challenge ensuring our market was ready to trade when the markets reopened.

Now, however, we have an opportunity to look back and consider not only what went well but also what could have been done differently or better. Through such critical analysis, the industry can become stronger and send a clear signal of the commitment we have all made to building the world's most robust equity markets. My testimony has some specific recommendations based on our experience.

While contingency planning can always improve our response to unforeseen events, perhaps the best plan that we can undertake is to foster competition between markets and enjoy the beneficial effects that produces. You know, if you fill up your tank at Texaco in the morning and the Texaco station goes belly up, you can always fill up at AMOCO across the street or Exxon down the block.

So, too, if Island has a system problem, all of our customers take about a second to start shunting all their orders to other people at this table. That is a very painful process for me, but is great for the customer; and just like the mythical hydra, a many-headed market structure where people have multiple connections to multiple competing marketplaces is the best way to insert continuity for the investor.

If we can all agree that strong competition between markets will ultimately provide us with the most efficient market model, the question becomes, what changes, if any, are necessary to ensure varied and vibrant competition between markets?

77-119 02 - 2

In Island's view, we should eliminate any barriers that inhibit fair competition between electronic and traditional markets. Currently, there are two main market structure changes that must be immediately pursued to ensure such fair competition. First, ECNs must be permitted to freely disseminate their market data to investors without sacrificing those very qualities that make ECNs such compelling alternatives in traditional marketplaces. Second, ECNs must have alternatives to Nasdaq in which to operate.

On the first point, Island's quotation data cannot be included in the consolidated tape mentioned earlier by the chairman and disseminated by market data vendors, because only markets that participate in the so-called intermarket trading system, or ITS, are permitted to have their quotation data disseminated in the consolidated tape. Unfortunately, Island cannot participate in the intermarket trading system because having such participation would undermine the very advantages of speed and certainty of execution upon which Island has built its business. I think this would put at risk the whole segment of trading and price discovery that have been created since Island began trading New York Stock Exchange, Nasdaq and AMEX-listed stocks.

I think the best way to conceptualize this ITS issue is to answer the following question: Would you take a $50 guaranteed seat to the World Series; or would you risk that for the off chance that you could get a $49 ticket, but not be sure that you could make other arrangements if you didn't get the ticket, or that you might even get the ticket the day after the game was played and end up not even going to the game. Speaking for myself, I will always take the guaranteed ticket and a chance to actually go to the game.

Island has had tremendous success in the last year trading selected listed stocks. In fact, in the largest AMEX- or New York Stock Exchange-listed security, the QQQ or Nasdaq 100 tracking stock, Island has now taken over 30 percent of the shares traded in the stock and has actually become the largest single marketplace. In fact, on these points, Nasdaq's Wick Simmons made these points in a December 4 letter to SEC Chairman Harvey Pitt.

I quote, "Because ITS orders require a minimum life of 60 seconds, it would greatly frustrate those investors and market professionals who wish to immediately lock in an execution. We are concerned that such a result would drive liquidity from U.S. Markets, producing a net loss for our Nation's economy and for its investors." And I will submit that letter for the official record.

The second point that I would like to make is that Nasdaq is the only market today that is not required to maintain strict time price priority through a so-called central limit order book on the exchange. The absence of a central limit order book makes Nasdaq the only place the ECNs can operate autonomously and efficiently.

To that end, the Cincinnati Stock Exchange recently filed a rule-making its central book and priority rules voluntary. In effect, Cincinnati filed a rule requesting to operate in the exact manner as Nasdaq, a market that competes with them in the same securities for the same customers. Island believes that quick approval of the Cincinnati Stock Exchange's filing will promote competition between markets strengthening our securities markets.

In conclusion, Mr. Chairman, I appreciate the subcommittee's interest in the Island's perspective on the events of September 11 and their implications for the markets; and I hope that the members of the subcommittee agree that in the wake of these events, we should redouble our efforts to create the strongest, most efficient market structure possible. I hope we can work together in the future on implementing both of Island's recommendations for fostering competition and thereby strengthen our equity markets.

[The prepared statement of Matthew Andresen follows:]

PREPARED STATEMENT OF MATTHEW ANDRESEN, CHIEF EXECUTIVE OFFICER, THE ISLAND ECN

Chairman Stearns and Members of the Subcommittee: I commend the Chairman and the Members of the Sub-committee for holding these hearings concerning the tragic events of September 11th and the role of Electronic Communications Networks in our securities markets.

I am Matthew Andresen, Chief Executive Officer of The Island ECN ("Island"). Island is an electronic marketplace that enables market professionals to display and match limit orders for stocks and other securities. Island's proprietary technology allows it to offer this low cost, rapid and reliable order display and matching service to a network of approximately 700 broker-dealers representing a diverse array of market participants. On an average day, Island will trade over 460 million shares—approximately one in every 5 trades on Nasdaq. Through November, Island has traded over 81.9 billion shares worth almost $2.5 trillion during 2001. Indeed, recent analyst reports indicate that Island is now the largest ECN in the world today.

Mr. Chairman, unquestionably, the events of September 11 had a profound impact on our Nation, not just our securities markets. Those of us in the New York financial community have felt an additional obligation to prove to the world the strength and resilience of America's stock markets. As members of New York City's downtown financial community, every Island employee joins with me in expressing our sadness for the events of that day and the loss of lives of innocent people, some of who were close to me. I am certain nobody will ever forget that day and how it changed our lives.

I am proud that despite the initial shock and even the fear of further attacks, Island rebounded quickly. From the employees who were on the roof cleaning out air conditioning ducts while burning soot blackened the whole of Wall Street to the team of technical staff that literally worked around the clock to reestablish connectivity to our subscribers, the Island staff rose to the challenge, ensuring our market was ready to trade when the markets reopened. The efforts at Island were replicated throughout the industry. Thanks to all the hard work, the contingency planning, the leadership of the Securities and Exchange Commission, and the sheer determination of the entire industry, the markets re-opened without incident on September 17.

Now more than three months later, we have an opportunity to look back and consider not only what went well, but what could have been done differently or better. Through such critical analysis, the industry can become stronger and send a clear signal of the commitment we have all made to building the world's most robust equity markets.

I believe that the industry was generally as prepared as possible for the events of September 11. Though we have already discussed our recommendations with the Securities and Exchange Commission, I would like to take a moment to touch on a few observations based on Island's experience:

1) *Back-up connectivity*—I think the entire industry learned the importance of maintaining connectivity not only to primary data centers but also back-up sites. Prior to September 11, some companies did not test or even maintain connectivity to Island's back-up facility. In the subsequent months, nearly all of our subscribers secured redundant connections to Island.
2) *Contingency planning*—The entire industry realized the importance of maintaining contingency plans. I am proud of how Island was able to re-establish its entire operation in our New Jersey back-up site within a day.
3) *Inter-market Coordination*—The SEC did an excellent job in keeping market participants informed. The SEC, however, may want to re-consider the composition of some of the emergency working groups. For example, although Island is approximately the same size as all the regional markets combined, Island was not asked to directly participate in many post September 11 meetings. Perhaps the

SEC should consider basing future working groups on relative market size rather than on regulatory designation.

While contingency planning can always improve our response to unforeseen events, perhaps the best planning we can undertake is to foster competition between markets and the beneficial effects it produces. Though some commentators, such as the Wall Street Journal, wrote articles suggesting that centralized physical markets are out-dated, Island does not believe that to be the case. While electronic markets may be less vulnerable to physical attacks, the extent to which centralized physical markets continue to attract orders demonstrates their continued necessity. Accordingly, Island would not support any initiative that would essentially dictate the structure of marketplaces. Instead, the future of the markets should be determined through competition. If all can agree that strong competition between markets will ultimately provide us with the most efficient market model, the question becomes "what changes, if any, are necessary to ensure fair yet vibrant competition between markets."

In Island's view, we should eliminate any barriers that inhibit fair competition between electronic and traditional markets. Currently, there are two main market structure changes that must be immediately pursued to ensure such fair competition. First, ECNs must be permitted to freely disseminate their market data to investors without sacrificing the very qualities that make ECNs compelling alternatives to traditional markets. Second, since all markets are competing in the same securities for the same customers, all markets must be permitted to operate under the same ground rules in the same manner. Let me more fully explain each one of these recommendations.

With respect to the first recommendation, many Sub-committee members are aware of the fact that while ECNs account for more than 50% of the transactions in Nasdaq listed securities, ECNs only account for approximately 5% of the transactions in listed securities. What accounts for this difference?

With respect to Nasdaq listed securities, ECN market data is permitted to be included in the quotation data disseminated to investors through the multitude of market data vendors. For example, if Island has the highest bid price in Cisco, which it often does, any investor or broker looking at information on Bloomberg, Reuters, Yahoo! or any other vendor service will see the high bid on Island. This transparency increases the likelihood that the order represented on Island will be executed and serves the best interests of investors.

In contrast, if Island has the highest bid or lowest offer price in the American Stock Exchange listed Nasdaq 100 tracking stock (known as "QQQ"), which Island has approximately 50% of the trading day, investors monitoring the same major vendor service will NOT see Island's best price. In fact, despite the fact that Island is regularly the largest market in the world for the QQQ, regularly matching more than 30% of the QQQ total shares traded on a given day, Island's market data is not included in the Consolidated Quote disseminated by vendor services. Island's exclusion from the Consolidated Quote prevents investors from receiving the best possible price.

Ironically, the National Market System, that was created to ensure that investors have access to all relevant market data, now prevents Island's market data in QQQ, and all other exchange listed securities, from being disseminated to investors. Further, the National Market System that was created to promote competition between markets now serves to inhibit competition from all-electronic markets.

Island's quotation data cannot be included in the Consolidated Quotation disseminated by market data vendors because only markets that participate in the Intermarket Trading System are permitted to have their quotation data disseminated in the official Consolidated Quote. Unfortunately, Island cannot participate in the Intermarket Trading System because such participation would undermine Island's advantages of speed and certainty of execution, destroying a whole segment of trading that has been created since Island began trading QQQ and preventing electronic markets from competing effectively in any exchange listed securities. The fact that Island has achieved this success with the severe handicap of not having its quotation data disseminated by market data vendors in the Consolidated Quote reflects the strong demand for a competitive alternative to traditional markets.

When it was created more than 20 years ago, the Intermarket Trading System was designed to ensure that investors received the best price irrespective of what market received the investor's order. The Intermarket Trading System, however, never contemplated that all-electronic agency markets would someday exist where users would demand responses within milliseconds of placing an order. Specifically, an Island subscriber utilizing the latest technology to implement a trading strategy that depends on millisecond responses should not be forced to send an order to a market that can respond up to 60 seconds later. Further, many market participants have reasonably concluded that the opportunity cost associated with waiting 60 seconds for the mere possibility of an execution exceeds the value of receiving an execution at a better price. If forced to always send the order to the best-advertised price, this new liquidity would vanish or go overseas, harming the competitive position of our Nation's markets. Nasdaq CEO Wick Simmons made these same points in a December 4 letter to SEC Chairman Harvey Pitt. "Because ITS orders require a minimum life of 60 seconds it would greatly frustrate those investors and market professionals who wish to immediately lock in an execution. We are concerned that such a result would drive liquidity from U.S. markets, producing a loss for the nation's economy and its investors."

A way to conceptualize the issue raised by the Intermarket Trading System is through the following question: "Would you take a guaranteed ticket to the World Series for $50 or take a chance on getting a ticket for $49 but not find out until it was too late to make other arrangements?" Speaking for myself, I would take the guaranteed ticket. Similarly, market participants should not be precluded by the rules governing the Intermarket Trading System from making the same choice when trading securities. Indeed, market participants trading Nasdaq listed securities are not required to always try to interact with the best-advertised price and Island is unaware of any widespread investor harm. Why, then, do we continue with a structure that we know prevents competition from electronic markets in exchange listed securities?

Island's second recommendation for promoting competition involves ensuring that all markets are able to compete on a level playing field. Currently, the Nasdaq market controls almost 100 percent of all transactions in securities listed on Nasdaq. Nasdaq is also the second largest market for exchange-listed securities. A key component of Nasdaq's monopoly in Nasdaq securities and success in listed securities is its unique regulatory structure that provides a regulatory advantage over other markets.

Specifically, Nasdaq is the only market that is not required to maintain strict price time priority through a central limit order book. The absence of a central limit order book makes Nasdaq the only place that ECNs can operate autonomously and efficiently. Nasdaq's monopoly on this type of market structure, however, restricts competition between markets. For example, no market can effectively compete with Nasdaq for hosting ECN trading unless they are permitted to change their rules regarding a central book and operate to match the rules of Nasdaq. To that end, the Cincinnati Stock Exchange recently filed a rule making its central book and priority rules voluntary. In effect, Cincinnati filed a rule requesting to operate in the same manner as Nasdaq, a market that competes with them in the same securities for the same customers.

Island believes that quick approval of the Cincinnati Stock Exchange's filing will promote competition between markets, strengthening our securities markets. Certainly, no market can continue to be permitted to use a regulatory inequity to maintain a monopoly over certain segments of market participants. I would hope that in the name of fostering fair competition that members of this Sub-committee would support Cincinnati's filing and urge expeditious review by the SEC.

CONCLUSION

In conclusion, Mr. Chairman, I appreciate the Subcommittee's interest in the Island's perspective on the events of September 11 and their implications for the market. I hope members of the Sub-committee agree that in the wake of these events we should redouble our efforts to create the strongest, most efficient market structure possible. This is done by ensuring fair and vibrant competition between markets, particularly competition between traditional markets and newer all-electronic markets such as ECNs. I hope that we can work together in the future on implementing both of Island's recommendations for fostering competition, thereby strengthening our nation's equity markets.

Mr. STEARNS. We thank the gentleman.

Ms. Kinney.

STATEMENT OF CATHERINE R. KINNEY

Ms. KINNEY. Good morning, Chairman Stearns, Congressman Towns and members of the subcommittee. My name is Catherine Kinney, and I am the Group Executive Vice President at the New York Stock Exchange. I, and our Chairman Richard Grasso, as well would like to thank you all for the opportunity to testify this morning on the industry's recovery following the attacks of September 11 and also the lessons that we have learned for electronic communications networks.

I wanted to cover a couple of specific areas this morning: First, as I said, thank you so much for having us here this morning and certainly to express the Exchange's sympathies for all who were involved in the attacks following September 11 and on September 11.

I also wanted to talk a little bit this morning about re-creating the environment, because I think some of the members of your committee started to do that this morning.

The devastation in downtown New York was enormous. The facilities of many of our broker-dealers, the major firms who supply order flow to the Exchange and who connect with many of the customers, representing both individuals and individuals themselves, were completely destroyed or rendered uninhabitable; and many of those firms had to relocate to New Jersey or other locations that had already been preestablished by those firms.

I think you all know and certainly have jurisdiction over telecommunications, so I know that you are very aware of Verizon's circumstances as well as Con Edison's circumstances. These were unprecedented for them. I think in our written testimony you will see we outlined that. But Verizon's central switching station was very heavily damaged, and actually they had no access to that building until September 14.

Con Edison lost 5 of its 7 feeder cables, also unprecedented. And those of you who have visited downtown Manhattan recently know that they had to run those cables through the streets and then cover them with wood and now we have nice bumps that you have to crawl over getting from street to street. But these are high voltage cables that obviously made our life easier and certainly gave us the opportunity to reopen our markets successfully on the morning of September 17.

But with the devastation and the infrastructure such as it was and the heroic effort of so many to reestablish the markets, I think it is very important to focus on what really happened following September 11.

Certainly the New York Stock Exchange's facilities were unaffected. I think there have been numerous reports and articles written that we didn't have power, we didn't have lots of things. The New York Stock Exchange facilities were unaffected, as well as two data sites unaffected. So if it were simply a matter of the New York Stock Exchange showing up to trade, that would not have been a problem for the Exchange. But the problem really was, as you will hear today and have already heard, our modern market model depends on connectivity; and that is not just connectivity to the market centers, but it is the connectivity to the customers and the member firms.

We described, and have described, our situation as a hub-and-spoke model, the Exchange being the hub and the spokes being the firms and their customers. As I said, if it were the matter of the hub only, I think most of the people here could have assembled something to trade. But it really was about a much larger population in the financial services infrastructure, which included the member firms.

Those member firms supply order flow to us, and the order flow is the lifeblood. It is the interaction of the buyers and the sellers each day which make the markets in our respective environments. But all of that order flow and getting that order flow to these respective models depends on connectivity. Without that, we are simply a hub with no spokes. The interruption of these connections meant that the buyers and sellers couldn't meet, whether it was on the New York Stock Exchange or any of these decentralized models or in cyberspace.

And really the debate, in our view, should not be about these central hubs, but rather about the issue of connectivity, about what we need to do with respect to restoring that connectivity and ensuring that in the future all of the connectivity and the alternatives and contingencies are well planned for.

I think that the two issues and two questions that faced all of us, and certainly those that participated in the variety of meetings that occurred following September 11 as we debated when to restart the markets, the two very important questions were the provision of liquidity and could we produce a fair and orderly pricing mechanism and pricing model that investors could rely on. And in doing that and in providing that liquidity, the two questions we asked, is it possible to engage in reliable price discovery if a significant source of order flow is cutoff, and is it fair to open those markets if a significant amount, or number of the buyers and sellers cannot reach the marketplace? So it is with those two questions that we began to look at restarting the markets following September 11.

As you know, meetings were conducted for 2 days by the SEC, or among the SEC, the Treasury, the Fed, the New York Stock Exchange, NASDAQ and all major market participants as well as the utilities. Clearly, the marketplaces and the member firms had issues in terms of this connectivity, but I can certainly say that in telecommunications, both Verizon and Con Ed played a major issue in terms of our ability to restart those markets, since they were not fully operational until at least Friday.

I think then, at the Thursday meeting following September 11, all of the participants were very concerned about a false start. I think it was said here this morning that everybody was very concerned about having orderly, reliable pricing and a market that would not start and fail again. I think that would have been something that both the American people and the financial services markets could not have withstood.

There was—we felt an opportunity both on Friday and over the weekend to continue to test the connectivity which was being restored. If you can imagine the number of firms relocating, having to repath, even as was described for Instinet here, repathing all of their connectivity to their customers as well as to the markets, all of that had to be retested and restarted. And the time that it took gave us that opportunity, the weekend was used to test the connectivity.

I think the results on Monday, September 17, spoke for themselves about the successful decisions that were made, that the work that went on for the days that followed September 11 was really the right work, and the focus was very apparent in the success of

September 17. We had a single record volume day of 2.3 billion shares. We had a record week, trading 10 billion shares. I don't think any of us who were at our respective sites on Wednesday, Thursday or Friday would have concluded or would have even thought we could have had that kind of volume.

Mr. STEARNS. Ms. Kinney, I will just have you sum up.

Ms. KINNEY. I think if we focus on lessons learned, every business is located somewhere. All of the trading platforms that will speak this morning are reliant on some physical plants but, as well, a very significant portion of us are reliant on our data processing equipment. We will have to continue to make significant investments to ensure that we have both the connectivity and reliable sites within a 24-hour basis in order to continue to trade. Thank you.

[The prepared statement of Catherine R. Kinney follows:]

PREPARED STATEMENT OF CATHERINE R. KINNEY, GROUP EXECUTIVE VICE PRESIDENT, NEW YORK STOCK EXCHANGE, INC.

Good morning Chairman Stearns, Congressman Towns, and members of the Subcommittee. My name is Catherine Kinney, and I am a Group Executive Vice President of the New York Stock Exchange ("NYSE"). I would like to thank the Committee for the opportunity to testify this morning on behalf of the NYSE, and our Chairman, Richard A. Grasso, on the industry's recovery from the attacks of September 11, and on the lessons from our experience for electronic communications networks ("ECN's").

Mr. Chairman as you know, the September 11th attacks devastated downtown New York City. Everyone in our industry lost a family member, a friend or a cherished coworker, and our thoughts and prayers remain with them.

The destruction of the World Trade Center complex resulted in tremendous collateral damage to the infrastructure of the entire area. Adjacent office buildings were rendered uninhabitable. Electricity, water and telephone service to much of the area was destroyed or disrupted. Verizon's central switching station was heavily damaged, and rendered inaccessible until the following Friday evening.

That the national securities markets were able to operate again just four business days later—at record volumes and without any systemic problems—is a tribute to an extraordinary partnership among the securities industry, the federal and local authorities, and the New York utilities. When the markets reopened, it was on an inclusive basis; every customer who wanted trade could access the market. The market did experience a 10% decline, but this adjustment was consistent with the European markets in the aftermath of the attacks. But the trading was fair and orderly, even in the face of record volume. During that first week of resumed trading, as the NYSE set new records for one day volume (over 2.3 billion shares on Monday September 17th) and first hour volume (660 million shares on Friday, September 21st), our market operated in a fair and orderly manner.

Mr. Chairman, order flow creates price discovery, and in the modern marketplace, order flow is dependent on connectivity. I will discuss these two factors in some detail, and then close with a brief mention description of the lessons we learned from the attacks.

"CONNECTIVITY", ORDER FLOW AND THE FINANCIAL MARKETS

Thankfully, NYSE people, facilities (including its our two remote data-processing sites), heavy infrastructure, networks, and trading systems and—thanks to diverse routing—most of our voice lines, were unaffected. So we had no need to move to an alternative site. Indeed, if we could operate our business solely by convening 3000 people on our trading floor, trading could have resumed immediately.

But today's markets, whether national exchanges, decentralized dealer markets, ECN's, or other alternative trading systems, depend on "connectivity", i.e., continuous access to the telecommunication systems that simultaneously link all market participants and provide the conduit for orders to interact and create markets. We live in an electronic age that permits participants from around the world to access, enhance, and benefit from, the unrivaled liquidity and order competition that takes place on the floor of the New York Stock Exchange.

Mr. Chairman, telecommunications connectivity, whether it be data or voice communication, is the lifeblood of our industry.

The NYSE, and our member firms which introduce the orders of some 85 million U.S. investors and millions of international investors, are linked through a complex, global communications and data delivery network. Through It is this electronic network, they place more than 90% of the orders that the NYSE executes, representing over 50% of our volume. The NYSE owns and operates fully redundant networks with no dependence on a third-party carrier, and is thus better positioned than others.

The hub (e.g. NYSE) and spokes (broker-dealers) of our securities markets represent a classic example of the "network effect" where connectivity is critical. Markets of all types deploy data networks to connect the broker-dealers transmitting orders to the markets. The broker-dealers, in turn, are dependent on networks for communications to their customers.

It is as simple as this: it takes the interaction of buyers and sellers to make markets. If, as on September 11th, something interrupts the connections of the broker-dealers to their customers and to the markets' networks, as occurred on September 11, the orders of buyers and sellers cannot meet—on the NYSE, on the trading floors of Nasdaq market-makers, or in cyberspace.

Although the market centers, (except for the American Stock Exchange) were intact, we soon knew that the industry had suffered a massive loss of connectivity. We and our government and industry partners understood that this loss of connectivity posed two questions that we had to answer in deciding when to reopen the markets.

First, from the perspective of the U.S. capital markets, and indeed, of the world, we asked, "Is it possible to engage in reliable price discovery if significant sources of order flow are cut off?" Second, from the perspective of those buyers and sellers who, on Wednesday, September 12, could not reach their brokers and the markets, we asked, "Is it fair to open the markets when a significant number of buyers and sellers are cut off?"

As you know, we answered both questions, "No."

RESTARTING THE MARKETS

The events of September 11th devastated the infrastructure of downtown Manhattan. Many firms had to relocate. Those whose physical plants were unaffected were deprived of the comprehensive network of communications, and the flow of information, that permits informed decisions, the placing of trades and the creation of deep pools of liquidity.

Nevertheless, four business days later, the equity markets were fully operational at record volumes and fair prices. Allow me to present a brief chronology of the events that made this possible.

During the six days following the attack, the NYSE, the other markets and our member firms, along with the city, state, and federal officials, Con Edison, Verizon and our central technology provider, the Securities Industry Automation Corporation (SIAC), worked continuously to restore and test the telecommunications infrastructure that ensures the connectivity of market participants. The task of recreating and rerouting downtown New York's telecommunications infrastructure, and ensuring that industry participants and end-users could access our data systems, was daunting. Many firms were forced to relocate. Communications routing systems had to be redirected or changed completely. Our staff assisted in this massive rerouting effort, testing every newly fashioned linkage. Virtually every aspect of that interconnected network—markets to firms and firms to customers, from the introduction of a trade to comparison and settlement—was verified.

Against the backdrop of the enormous task of relocating, rewiring and rerouting, on Wednesday, September 12, representatives of the SEC, Treasury and the Federal Reserve System, as well as of the three principal equities markets, met with the senior management of the major financial institutions. It was clear to all participants at that Wednesday's meeting that the physical devastation in downtown New York City precluded opening the next day.

It is hard to describe the extent of the damage to downtown New York's infrastructure. Verizon's switching station had been flooded, and was to remain completely inaccessible until Friday night. Five of Con Edison's seven feeder cables that run the downtown power grid were destroyed. Con Edison was forced to recreate this grid above ground, and run 135,000 volt cables down the street. Even those firms that had not suffered physical damage had no long-distance service, so they could not reach their customers.

At this Wednesday meeting, we agreed to reconvene in 24 hours to review the situation with a clearer understanding of when connectivity could be reestablished with the broker-dealer community and with customers. The decision to resume trading on Monday was made during a Thursday, September 13th, meeting of essentially the same group. We all knew that opening the equities markets without assuring connectivity could result in illiquid markets if buyers, sellers and their brokers could not access the market centers. We concluded that the resulting lack of liquidity would produce unprecedented volatility and suspect prices, leading to a crisis in investor confidence. We further determined that a premature or false start would do greater damage than delaying the opening of our markets until the next business day.

So we determined it was prudent to resume trading on Monday, September 17th. This gave the utilities a chance to restore services to broker-dealers and the 85 million American investors they serve. It permitted our member firms to complete their reconnection of data and voice communications to their customers and to us, and for us to test to ensure they had succeeded in reconnecting. The best evidence of our success in this impressive collective effort was the seamless resumption of trading on Monday, September 17—the most active day in our history. The NYSE handled a record 2.3 billion shares, twice our average volume. Our market was liquid. Every system worked. Every buyer and every seller had a way to reach our auction. No one was left out.

Indeed, thanks to the ubiquity of our fiber optics network, we even managed to establish the American Stock Exchange's equities market on our floor.

LESSONS LEARNED

What lessons have we taken from meeting this extraordinary challenge? What are we doing to be even better prepared in the future?

Planning for contingencies that would have seemed farfetched a year ago now seems prudent. As an initial response, we have made the investments necessary to make ready on shorter notice a fully equipped alternative-trading floor. This alternative facility could be operational within 24 hours if any future attack rendered our trading floor unusable.

Our constituent board of directors—CEO's of member firms, issuers and institutional investors, as well as leaders drawn from the public sector—has met twice since September 11. Each meeting has included a discussion of the nature of the threats that September 11, made so apparent and of how we should respond. My colleagues and I are engaged in a contingency planning process that looks beyond our current alternative floor, and takes into account the opportunity presented by our planned new trading facility across Broad Street, as well as by other possible sites.

Mr. Chairman, let me close by saying that, "Every business is located somewhere." Every market center, be it a floor-based exchange, a decentralized dealer market or an ECN, relies on physical locations for personnel and data processing equipment.

Again, I want to thank you, Mr. Chairman, for the opportunity to appear before you today.

Mr. STEARNS. I thank you.

Just to remind the panel of witnesses, generally the opening statements are about 5 minutes. We are allowing you to go over 5 minutes, but certainly want to keep it below 10 minutes. So we would appreciate it as close to 5—that would be very helpful.

Mr. Bang.

STATEMENT OF KIM BANG

Mr. BANG. Mr. Chairman and members of the subcommittee, my name is Kim Bang.

Mr. STEARNS. Mr. Bang, just move that a little closer to your mouth.

Mr. BANG. I am President of Bloomberg Tradebook. I am pleased to testify on behalf of Bloomberg Tradebook regarding ECNs in the wake of September 11.

Bloomberg Tradebook is an electronic agency broker. We count among our clients many of the Nation's largest institutional investors, who themselves represents millions of individual investors. Bloomberg Tradebook specializes in providing innovative tools that enable our clients to step unobtrusively into the electronic crowd of the national market system to find liquidity for themselves and, in the process thereby, also providing liquidity for others. Our clients have rewarded our creativity and our service by entrusting us with their business, allowing us to regularly execute 200 million shares a day.

The reason ECNs now account for more than 46 percent of the reported share volume on Nasdaq is simple. ECNs are a market solution to investor demand. By providing a combination of neutrality, transparency, fairness and innovation, investors are now empowered with the direct access to liquidity in the national market system.

By definition, we are agency brokers. We take no position for our own accounts, and thus, we are neutral in the marketplace. We exist to service our customers' execution needs who want to buy and sell shares.

Over the past year, we estimate that we have saved our clients in excess of $1 billion dollars in transactional costs alone. Like market-makers, we maintain an electronic book of our customers' bids and offers. But unlike market-makers, we publish our entire book of quoted prices electronically for our customers to see. Indeed, unlike some of the ECN competitors, we employ an open architectural platform designed to route our customer orders to the best available price even when that means the price does not exist in the Bloomberg Tradebook system.

Without a government-sponsored monopoly to rely on, the market commands that ECNs compete and innovate. For example, at our inception in 1996, Bloomberg Tradebook instituted the concept of a reserve. A reserve is a process that controls the release of an order into the marketplace, enabling clients to trade large orders more efficiently. Like all innovations, the reserve gave us a leg up initially on our competitors, but only for a brief period of time. It has since become an industry standard, and today nobody would introduce a system without it. Any edge that we gain is a momentary one, and we are forced to continue to innovate, and we have done so continuously over the past 5 years.

When the Senate Banking Committee held a hearing in the last Congress, exploring the role of ECNs, Frank Zarb, then chairman of the National Association of Security Dealers, stated that, "I guess I sum up the answer as to why we have ECNs as the fact that the national stock exchanges, and I am not only talking about ours, but the exchanges around the world haven't been keeping pace with the needs of the market."

Mr. Chairman, it is worth pondering why the stock exchanges didn't keep pace with investor demands, as Mr. Zarb stated. We would submit that a government-sponsored monopoly ultimately cannot provide those innovative ideas and customer solutions of the best ECNs precisely because they are government-sponsored monopolies.

At present, most SROs are nonprofit organizations. The Nasdaq, however—the NASD, however, has largely completed its privatization of Nasdaq, and it may well be that other privatizations are going to follow. For-profit exchanges will have powerful incentives to leverage their existing government-sponsored advantages to gain an unfair advantage in current competitive markets. They will have incentive to keep pace with market innovators not by moving forward themselves, but by slowing down other market participants that tend to centralize order flow.

As to the specific tragedy of September 11, the financial service industry was Ground Zero of the attack on America. All of us in this industry have suffered enormous losses. At Bloomberg, the steps we have taken include the immediate allocation of substantial sums of money to provide free office space and support for our clients, including phones, computers, Bloomberg terminals and other necessities needed, in fact, more than 1,200 displaced financial workers since the tragedy.

Many of them have lost friends and colleagues, and we at Bloomberg, ourselves, have lost three people. Yet our clients and everyone involved inspire us with their commitments to get back to work and their display of the extraordinary strength in the human spirit.

As a technological matter, Bloomberg could have traded on the afternoon of September 11. That statement does not, however, imply any criticism of the collective and difficult decision to close the markets during this unprecedented crisis. As to the long-term public policy lessons to be gleaned from this tragedy, I have observed there has been a debate over the past few years over whether public policy should favor a more decentralized market structure or whether public policy should encourage a more centralized market structure as often advocated by the exchanges. I believe, if there is anything approximating a level playing field, market forces will drive toward decentralization. Clearly, September 11 underscored the wisdom of moving in that direction.

In conclusion, billions of transaction costs have been saved by ECNs in the Nasdaq marketplace. Investors in the New York Stock Exchange's listed market should be permitted an opportunity to enjoy the same benefits. As the exchanges that have traditionally functioned as public utilities become for-profit entities, investors will suffer if the exchanges succeed in leveraging their existing government-sponsored monopolies into currently competitive arenas. We should opt instead for a continuation of tremendous progress that has been made over the past 5 years in the Nasdaq marketplace. We should attempt to allow similar competition and innovation to flourish in the listed market, thus preserving America's status as the world's premier capital market.

Thank you.

[The prepared statement of Kim Bang follows:]

PREPARED STATEMENT OF KIM BANG, PRESIDENT, BLOOMBERG TRADEBOOK LLC

Introduction. Mr. Chairman and Members of the Subcommittee. My name is Kim Bang, and I am pleased to testify on behalf of Bloomberg Tradebook regarding ECNs in the wake of September 11th.

Bloomberg Tradebook is owned by Bloomberg L.P. and is located in New York City. Bloomberg L.P. provides multimedia, analytical and news services to more than 150,000 terminals used by 350,000 financial professionals in 100 countries worldwide. Bloomberg tracks more than 135,000 equity securities in 85 countries, more than 50,000 companies trading on 82 exchanges and more than 406,000 corporate bonds. Bloomberg News is syndicated in over 350 newspapers, and on 550 radio and television stations worldwide. Bloomberg publishes seven magazines, as well as books on financial subjects for the investment professional and non-professional reader.

Bloomberg Tradebook is an electronic agency broker serving institutions and other broker-dealers. We count among our clients many of the nation's largest institutional investors. Bloomberg Tradebook specializes in providing innovative tools that allow our clients to step unobtrusively into the electronic "crowd" of the new national market system to find liquidity for themselves and, in the process, provide it for others. Our clients have rewarded our creativity and our service by trusting us with their business, allowing us to regularly match in excess of 200 million shares a day.

What are ECNs? Before analyzing the effects of September 11th on ECNs, we must first explore what ECNs are and how they came into existence. There has been an enormous growth in ECNs over the past five years. That growth has been made possible by the issuance in 1996 of the SEC's Order Handling Rules. These rules—aimed primarily at exchange specialists and Over-the-Counter market makers—were designed to promote market transparency.

A few years ago, the SEC took a number of steps to reform the markets—starting with directing the reorganization of the governance structure of the NASD. On a limited number of critical committees, the NASD was directed to provide for significantly greater involvement by representatives of the public and NASD constituencies other than market makers. I am privileged to serve on one of the key committees cited in the order—the Quality of Markets Committee. The express purpose of the SEC in directing these changes was to alter the perspectives of the NASD and infuse the NASD with a greater sense of objectivity and impartiality.

It is often remarked that sunlight is the best disinfectant, and, indeed, the increased transparency promoted by the SEC's Order Handling Rules and the subsequent integration of ECNs into the national quotation montage narrowed Nasdaq spreads by nearly 30% in a year. While the complete list of reforms ordered by the SEC to promote transparency is long and varied, all of these changes, including the promulgation of the Order Handling Rules, were animated by the same underlying principle—namely that the sunlight produces the most honest and efficient markets.

ECNs—A Market Solution to a Market Problem. A regulatory regime that encourages transparency was a necessary, but not sufficient, precondition to the growth of ECNs. The reason ECNs account now for more than 46% of the reported share volume of Nasdaq is simple—ECNs are a market solution to a market problem.

ECNs are distinguished by four characteristics—neutrality, transparency, fairness and innovation.

Neutrality? By definition we are agency brokers and take no position for our own accounts. Thus we are neutral in the marketplace and exist only to serve our customers' need to buy or sell shares.

Transparency? Like market makers, we maintain an electronic book of our customers' bids and offers. But unlike market makers we publish our entire book of quoted prices electronically for all our customers to see. Indeed, unlike some of our ECN competitors, we take advantage of this transparency to route our customers to the best available price, even if it is outside of Bloomberg Tradebook.

Fairness? ECNs are required by SEC rules to respond immediately—and I mean immediately—to orders in the order they are received, whether they come from our best customers or from our competitors. That's probably the highest standard of fairness in the industry.

Innovation? Without a government-sponsored monopoly to rely on, the market demands that ECNs innovate. For example, at its inception in 1996, Bloomberg Tradebook introduced the concept of "Reserve" to the U.S. equity markets. "Reserve" is a process that controls the release of orders into the market, enabling clients to trade large orders more efficiently. Like all innovations, the "Reserve" gave us a leg up on our competitors for a brief period of time. It has since become the industry standard. Today no one would introduce a system without it. Any edge we gain is a momentary one—and we are forced to continue to innovate. We have done so continuously in the past five years.

Along with neutrality, transparency, fairness and innovation, add lots of enthusiasm and creativity from people passionately devoted to serving their customers and you have a picture of who we are and why we exist.

When the Senate Banking Committee held a hearing in the last Congress exploring the role of ECNs, Frank Zarb, then-Chairman of the National Association of Securities Dealers, stated that "... I guess I sum up the answer as to why we have ECNs as the fact that the national stock exchanges, and I'm not only talking about

77-119 02 - 3

ours, but the exchanges around the world haven't been keeping pace with the needs of the market."

Mr. Zarb is an accomplished leader in business and public service. Investors are fortunate to have had the benefit of his leadership, but I respectfully submit that the reason ECNs exist is not only because of what national stock exchanges failed to do, but also because of what we innovating broker-dealers have done, in the heat of competition. Mr. Chairman, it's worth pondering why the stock exchanges didn't keep pace, as Mr. Zarb stated. We would submit that a government-sponsored monopoly ultimately cannot provide the innovative ideas and customer service of the best ECNs precisely because they *are* a government-sponsored monopoly. To spur future innovation, I'd rather place my faith in the NASD's members—the marketplace of competing broker-dealers.

The Current Challenge. At present, most SROs are non-profit organizations. The NASD, however, has largely completed its privatization of Nasdaq and it may well be that other privatizations will follow. Under the cover of a non-transparent bureaucracy, non-profit SROs have exploited the opportunity to subsidize their other costs (e.g. costs of market operation, market regulation, market surveillance, member regulation) through market information fees. For all SROs, the incentive will be strong to exploit this government-sponsored monopoly over market data by charging excessive rates for market data and using those monopoly rents to subsidize their competitive businesses. Indeed, shareholders of these now for-profit entities will effectively demand that market data charges remain excessive.

Along with its market data monopoly, Nasdaq will also have a powerful incentive to leverage its trade execution monopoly to the detriment of consumers, investors and the markets. Currently, there is no real alternative to Nasdaq's monopoly with respect to the execution of market-maker quotations/orders in securities traded via Nasdaq. Through a series of developments, starting with the inauguration of the Small Order Execution System ("SOES") in the 1980's and progressing through the development of SuperSOES and SuperMontage, Nasdaq has evolved from a decentralized, quotation and telephone-based system into a screen-based, electronic communications network embodying an electronic limit order book.

In theory, NASD members can bypass SuperSOES through private wire connections between a market maker and a customer or dealer. In reality, however, that means of avoiding SuperSOES is not on an equal competitive footing with the use of SuperSOES. Orders transmitted through SuperSOES impose obligations on the market maker to execute against its published quotation. Those obligations are not replicated by private wire connections.

Only Nasdaq has the monopolistic power to execute transactions against market makers quotations. Individual market participants do not have the market power to replicate that obligation through private contractual arrangements or other private ordering.

The same issue is raised by the recent approval by the Nasdaq Board of a pricing proposal that will charge excessive connectivity fees for routing trades outside of Nasdaq. This proposal was approved by the Nasdaq Board despite having been rejected by the Quality of Markets Committee by a 14 to 6 vote. If the best price for a stock exists outside of Nasdaq, consumers should be able to avail themselves of that price without the obstacle of a connectivity fee that bears no relationship to actual costs incurred in consummating the transaction.

The NASD currently is putting together a proposal for a new Alternative Display Facility ("ADF"). The ADF is intended to provide a competitive alternative to the Nasdaq SuperMontage/SuperSOES facility. It remains to be seen, however, how effective an alternative the ADF will be and whether it will be adequately funded. As it is, Nasdaq has taken unto itself the enterprise value of its market system, which the NASD developed over 30 years. Nasdaq embodies both a quotation facility and an execution/clearance facility, which the ADF is not intended to provide. It may be that the ADF will nevertheless be a preferred venue, but that will eventuate only if it is allowed to compete on an equal footing with Nasdaq. Exclusionary and anti-competitive elements in the SuperMontage/SuperSOES combination should be revised to provide that equal footing.

Currently, there is no "glue" in the proposed ADF. In the absence of mandated intermarket connections between Nasdaq and the ADF, the ADF may become a ghost town.

For-profit exchanges will have powerful incentives to leverage their existing government-sponsored monopolies to gain an unfair advantage in currently competitive markets. They'll have incentives to "keep pace" with market innovators not by moving forward themselves, but by slowing down all market participants and centralizing order flow.

If that occurs, consumers, investors and the markets themselves will be denied the benefits of competition. Everyone loses if exchanges—comfortable as government-sponsored monopolies—fail to innovate, leaving American markets vulnerable to offshore competitions.

ECNs—Consumers and Investors Benefit. So who has benefited from the existence of ECNs? For one, small retail customers who, for the first time, have gained direct unfettered access to the liquidity of institutional order flow represented directly in the market. Institutional investors who, for the first time, are able to find liquidity for their orders by interacting directly with small order flow. All investors who have seen the speed and fairness of their executions improve, as ECNs have raised the standard for all broker-dealers. Even traders not participating in ECNs benefit from our depth, liquidity and immediacy each time they hit an ECN bid or take an ECN offer.

Who Hasn't Benefited from ECNs? Useful linkages have yet to be developed for the New York Stock Exchange listed market. As a result, investors in that market have yet to reap the full benefits of the competition provided by ECNs. While the SEC has allowed ECNs access to the Intermarket Trading System (ITS) through Nasdaq, this is not sufficient. ITS remains crippled both by its technological ineffectiveness and an unworkable governance structure that makes any movement nearly impossible. As stated above, the same issue will arise if there is not an effective linkage between Nasdaq and the proposed ADF, without which a viable third market in Nasdaq securities likely will be impossible.

Government-sponsored market centers like the Nasdaq Stock Market and the New York Stock Exchange can either make ECN transparency available to the entire national market system or reduce transparency by seeking to block ECN display linkages. Clearly the NYSE has historically had no interest in encouraging linkages that would make ECNs players in the listed market.

ECNs in the Wake of September 11th. The financial services industry was ground zero of the attack on America. All of us in this industry have suffered enormous loss. At Bloomberg, we have been privileged to provide free office space and support—phones, computers, Bloomberg terminals, and whatever else is needed—to more than 1,200 displaced financial workers since the tragedy. Many of them have lost friends and colleagues—as we at Bloomberg did. Yet they inspire us with their commitment to getting back to work and their display of the extraordinary strength of the human spirit. They convince us that, although the terrorists have inflicted profound losses, they have not diminished our resolve. That resolution, that spirit of cooperation and sacrifice has animated the incredible ongoing efforts by so many in both the public and private sectors to resurrect our securities markets.

It is very difficult to think of September 11th in terms of its policy ramifications. It is clear, however, that all industries must go through the painful process of analyzing whether there are applicable lessons to be gleaned from this tragedy. We believe there are.

While I've described how the Order Handling Rules and the market's demand for these services made the growth of ECNs in the Nasdaq market possible, the third critical component is, of course, the advent of modern telecommunications and computing technology. This technology has facilitated a volume and speed of trading that would have been unimaginable not so long ago.

Technology makes possible a market structure that wouldn't have previously been possible. That has spawned a debate over the past few years over whether public policy should favor a more decentralized market structure, or whether public policy should encourage centralization as advocated often by the exchanges.

This argument has manifested itself in a number of different ways. A few years ago, proponents of centralization urged support for a time priority Central Limit Order Book (CLOB) to deal with the alleged "problem" of market fragmentation. The notion behind the CLOB was that, by centralizing orders in one place, a single "black box", maximum order interaction and perhaps better prices might be achieved.

While the CLOB was ultimately rejected, the previously described interaction of SuperSOES and SuperMontage within Nasdaq represent the same effort to centralize. The recent Nasdaq pricing proposal, which would clearly discourage execution of trades outside of Nasdaq—even if the best price for a stock were being offered outside of Nasdaq—is simply the latest manifestation of this urge towards centralization. As exchanges contemplate for-profit futures, this urge to centralize order flow and execution will grow more pronounced. This emphasizes the need for a functional, fully competitive ADF as a means to mitigate the anti-competitive effects of Nasdaq's market scheme. It may well be that additional remedial measures will be needed. The continued vigilance of the Congress and the SEC will be essential as these developments unfold.

As the growth of ECNs illustrates, modern technology allows the advantages of maximum order interaction without the downside of centralization. State-of-the-art telecommunications systems like the Internet don't rely on a single monopoly channel—rather they rely on networked webs of multiple competing and redundant linkages. Why should the securities markets work differently?

In addition, centralized systems are resistant to change. The innovations that ECNs have brought to the market would not have occurred under more centralized systems.

A centralized system also provides the significant downside of a central point of failure. Those of us who deal regularly with Nasdaq's SelectNet system know only too well how cumbersome and inefficient a centralized system can be. Like SelectNet, the ITS system is conceded even by the sympathetic to be technologically outmoded, with a bureaucracy that thwarts change. Why make those failed systems the model?

Bloomberg Tradebook, as well as others in our industry, has expressed concerns for years about the problem of a single point of failure posed by a centralized system. The tragedy of September 11th underscores that concern.

Conclusion. I thank the Committee for the opportunity to describe the regulatory structure, investor demand and the technological advances that have made possible the enormous growth of ECNs in the Nasdaq market. The neutrality, transparency, fairness and innovation we collectively bring to the Nasdaq market have dramatically increased efficiency on Wall Street, redounding to the benefit of Main Street and the economy. Investors in the New York Stock Exchange listed market should be permitted an opportunity to enjoy the same benefits.

Historically not-for-profit exchanges are contemplating a for-profit future. As market players that have traditionally functioned as public utilities become for-profit entities, their goals, incentives and agendas radically change as well. Consumers and investors will suffer if exchanges succeed in leveraging their existing government-sponsored monopolies into currently competitive arenas. These efforts will increase centralization in a manner that is not only unnecessary given modern technology, but also economically ill advised and potentially perilous.

Mr. STEARNS. Thank you, Mr. Bang.

Mr. O'Hara.

STATEMENT OF KEVIN J.P. O'HARA

Mr. O'HARA. Good morning, Chairman Stearns, Congressman Towns and other distinguished members of the subcommittee. On behalf of Archipelago, I am pleased and honored to be with you this morning and thank the subcommittee for holding this hearing.

Archipelago was co-founded by Jerry Putnam, our Chairman and CEO, and software developers MarrGwen and Stuart Townsend in late 1996. From January 20, 1997, the day Archipelago executed its first order as 1 of the 4 original "qualified" ECNs, its operating business has grown to average over 200 million shares per day, or roughly 10 percent of Nasdaq's overall volume, and 25 million shares per day of NYSE- and AMEX-listed volume.

In October of this year Archipelago was approved by the Securities and Exchange Commission to become the first fully open electronic national stock exchange. Through its business arrangement with the Pacific Stock Exchange, the Archipelago Exchange will launch early next year with its "best execution" business model at its core. The Archipelago Exchange, like its younger brother the Archipelago ECN, will route orders to superior prices if they exist outside the Archipelago system.

Further, only 3 weeks ago Archipelago and REDIBook, the two fastest growing ECNs, announced their intention to combine businesses. This merger of equals brings together two deep pools of liquidity into one fully integrated and innovative trading platform. The combined owners of Archipelago and REDIBook represent diverse investors from all walks of life, including institutional and retail brokers as well as professional trading houses and established Wall Street firms.

September 11 has compelled our Nation to fundamentally rethink risk—the risk of future attacks, the risk of providing global leadership and even the risk of an open society. In terms of our capital markets, the extreme concentration in lower Manhattan appears now to pose unsustainable geographic and economic risk. Indeed, the Wall Street Journal and the Washington Post both editorialize that the week-long trading hiatus from September 11 to September 17 was far too long, given our 21st century resources and wherewithal.

How do we manage the glaring risk exposed by September 11, and what does the future portend for our capital markets? An educated guess envisions a network of multiple competitive market centers linked by robust linkages which compete for business on the basis of price, product and service. A system of linked competitors is identical to the Internet model designed to provide redundancy and avert a single point of failure. It was precisely this decentralized model that proved unconditionally successful as a means of communication on September 11.

This notion is not new to the financial markets. In 1975, Congress laid out the road map for a national market system of informationally linked competing exchanges. The question, therefore, is not whether linkages exist or if electronic facilities such as ECNs and regional exchanges located outside lower Manhattan could have shouldered the burden of trading on September 12, 13 or 14. If called upon, we believe we could have answered the bell. Rather, the query is whether our industry is prepared to move to embrace a less centralized model and thereby eliminate the risk of shutting our markets down in the face of the unthinkable.

In our quest to manage risk of the unthinkable, perhaps we can draw lessons from the thinkable. An overused criticism of alternative markets such as ECNs disputes that their efficiencies are only available when times are good and the market is going up but are nowhere to be found when markets are stressed or, in industry parlance, are "cratering." The argument continues that only the anointed specialists or market-makers would be there to "catch the falling knife" by buying in the face of extreme selling.

However, empirical evidence to date erodes the "falling knife" critique. This is evidenced by data reflected in the display chart off to your right there, Enron trading from November 28, 2001, which plots the price action of Enron, the beleaguered energy giant, on November 28, 2001.

On that day, Standard & Poors announced a downgrade of Enron's debt to junk status, which sent the stock into another leg down in its free-fall. Minutes before the announcement, the New York Stock Exchange halted trading at 10:58 a.m. Due to a, quote, order imbalance. In other words, there were more sellers of Enron than buyers, and the "knife" was too sharp for the specialist to catch. Note that unlike a regulatory halt, which is marketwide, a halt for an order imbalance or operational failure does not impact the ability of other markets to trade.

While the New York Stock Exchange specialist responsible for trading Enron shut down his post over the next half hour, ECNs and other alternative venues traded over 10 million shares of Enron as the stock went from $2.60 to $1.10. At 11:27 a.m., the New York Stock Exchange specialist resumed trading in Enron at prices discovered by these alternative markets.

The upshot: No single entity, be it exchange, specialist or market-maker, can go it alone to catch a falling knife. Efficient price discovery is the product of the entire marketplace, including ECNs and alternative exchanges.

Finally, before I conclude, I would be remiss in not commending Congress, the SEC, and public advocates for supporting the conversion of our equity markets to decimal pricing. In particular, the Commerce Committee was a critical catalyst for this positive change that, to date, has narrowed effective spreads in the most liquid stocks on Nasdaq and the New York Stock Exchange by an average of 50 percent and 15 percent respectively. This fundamental change has directly led to enormous reductions in trading costs and, importantly, puts tens of millions of dollars back in the pockets of investors.

Thank you for your steadfast perseverance.

[The prepared statement of Kevin J.P. O'Hara follows:]

PREPARED STATEMENT OF KEVIN J.P. O'HARA, GENERAL COUNSEL, ARCHIPELAGO HOLDINGS, L.L.C.

I. INTRODUCTION

Good morning Chairman Stearns, Congressman Towns and other distinguished members of the Subcommittee. On behalf of Archipelago Holdings, L.L.C. ("Archipelago"), I am pleased and honored to be with you this morning, and commend the Subcommittee for holding this hearing in the wake of the September 11 attacks.

Archipelago was co-founded by Jerry Putnam, our Chairman and CEO, in late 1996, and software developers MarrGwen and Stuart Townsend. From January 20, 1997, the day Archipelago executed its first order as one of the four original "qualified" Electronic Communication Networks ("ECN"), its current operating business has grown to average over 200 million shares per day, or roughly 10% of Nasdaq's overall volume, and 25 million shares per day of NYSE- and Amex-listed volume. Today, Archipelago is the only ECN to reflect a quote in the National Market System for listed securities, such as AOL and IBM.

In October of this year, after working side-by-side with the dedicated staff of the Securities and Exchange Commission ("SEC" or "Commission"), Archipelago was approved by the Commission to become the first fully open electronic national stock exchange. Through its business arrangement with the Pacific Stock Exchange, who will serve as its regulator, the Archipelago Exchange will launch early next year with its "best execution" business model at its core. The Archipelago Exchange—like its younger brother the Archipelago ECN—will route orders to superior prices if they exist outside of the Archipelago system. The Archipelago Exchange will be fully integrated into the National Market System and will compete toe-to-toe with the NYSE, Nasdaq, and Amex.

Further, only three weeks ago, Archipelago and REDIBook, the two fastest growing ECNs, announced their intention to combine businesses. This merger of equals brings together two deep pools of liquidity into one fully integrated and innovative trading platform. The combined owners of Archipelago and REDIBook represent diverse investors from all walks of life, including institutional and retail brokers as well as professional trading houses and established Wall Street firms. They include American Century Funds, Charles Schwab, Goldman Sachs, Credit Suisse First Boston, E*Trade, Fidelity Investments, BNP Cooper Neff, J.P. Morgan Chase, Lehman Brothers, Merrill Lynch, Fleet Securities, Pershing, Spear Leeds & Kellogg, TD Waterhouse and CNBC. Archipelago and REDIBook look forward to closing their transaction and are excited to getting down to work in delivering value to investors and competing effectively against traditional exchanges and marketplaces.

II. THE UNTHINKABLE: SEPTEMBER 11, 2001

In preparing for this hearing, I had the opportunity to peruse the statements of many distinguished administration officials, regulators, and industry representatives who have testified before congressional committees. I will not attempt to match their perspective or ability to articulate the unthinkable events and repercussions of September 11, but will only highlight the recent words of SEC Chairman Harvey Pitt: "The events surrounding this meeting are both a cause for grieving and a cause for giving thanks." At Archipelago, we were blessed by good fortune unlike some of friends and neighbors in the industry. While shaken, none of our employees or their families were killed or injured in the attacks. Although headquartered in Chicago, Archipelago maintains its second largest office, complete with a backup data center, just a stone's throw from the NYSE. After assuring the safety of our New York employees, we began the process of restoring power to our New York office with round-the-clock help by the NYPD, the NYFD, ConEd, the Army Corps of Engineers, and FEMA. Like other marketplaces, the SEC asked us whether we would be ready to go in the immediate days following September 11. And, we responded, "we are," though we deferred to the SEC, Congress and the Administration as to when we should begin trading again. On Monday, September 17, I am happy to report that Archipelago joined other securities markets in a fully successful reopening. I might add that we, like the rest of the financial industry, owe a particular debt of gratitude to, among others, Mayor Guliani, SEC Chairman Harvey Pitt, Congressman Fossella, Senator Schumer, Congressman Towns and the staff of the Nasdaq and NYSE for successfully reopening our markets on that Monday.

Yet, despite the Herculean efforts and heroic actions, September 11 has compelled our nation to fundamentally rethink "risk": the risk of future attacks, the risk of providing global leadership and, even, the risk of an open society. Once assessed, the rational response is how best to manage these risks. In terms of our capital markets, the extreme concentration in Lower Manhattan appears now to pose unsustainable geographic and economic risk. Indeed, *The Wall Street Journal* and *Washington Post* both editorialized that the weeklong hiatus was far too long for our markets to be closed, given our 21st century resources and wherewithal.

III. THE RISK MANAGEMENT CHALLENGE: A COMPETITIVE RESPONSE

In the face of adversity, our nation is nothing if not resourceful and flexible and innovative. We quickly learn from our mistakes, and our robust financial system is no exception.

Archipelago has spent the last two years evolving from an ECN to a fully electronic national securities exchange. Others, such as many of my colleagues on the panel today, have blazed the same or similar trails. Though not yet complete, the aggregate effect of our innovations has been the metamorphosis from a floor-based, utility stock-trading model to an electronic, for-profit one.

How do we manage the glaring risks exposed by September 11, and what does the future portend for our capital markets? An educated guess envisions a network of multiple competitive market centers, linked by robust linkages, which compete for business on the basis of price, product, and service. A feature of "service" is certainly accessibility: some markets will offer a floor-based solution, with the advantages of "high touch" order handling, while others will offer screen-based and anonymous access, perhaps as a means to mitigate geographic risk. A system of linked competitors is identical to the Internet model, originally designed to provide redundancy and avert a single point of failure. It was precisely this decentralized model that proved unconditionally successful as a means of communication on September 11.

This notion is not new to financial markets. In 1975, for instance, Congress laid out the roadmap for a National Market System of informationally-linked competing exchanges, including the now-outdated Intermarket Trading System completed in 1980. The question, therefore, is not whether linkages exist, or if electronic facilities such as ECNs and regional exchanges located outside Lower Manhattan could have shouldered the burden of trading on September 12 or 13. If called upon, we believe we could have answered the bell. Rather, the proper query is whether our industry is prepared to move to embrace a less centralized model, and thereby eliminate the risk of shutting our markets down in the face of the unthinkable.

IV. ONE OF THE "THINKABLE": ENRON ON NOVEMBER 28

In our quest to manage risk of the unthinkable, perhaps we can draw lessons from the thinkable. An overused criticism of alternative markets (*e.g.*, ECNs, ATSs) disputes that their efficiencies are only available when times are good and the market is going up, but are no where to be found when markets are stressed or, in industry parlance, "cratering." The argument continues that only the anointed specialist or market makers would be there to "catch the falling knife" by buying in the face of extreme selling. This criticism speaks to the core hesitancy that some have with safeguarding our financial markets to those of us without 209-year operating histories. It goes something like this: "Newcomers can't be trusted in stressful times."

However, empirical evidence to date clearly erodes the "falling knife" critique. And, with the growth of alternative markets and continued technological progress, the "falling knife" critique loses credibility with each passing day. This is evidenced by the data reflected in Exhibit A (a copy of which is attached hereto), which plots the price action of Enron, the beleaguered energy giant, on November 28, 2001.

On that day, Standard & Poor's announced a downgrade of Enron's debt to junk status, which sent the stock into another leg down in its freefall. Minutes before the announcement, the NYSE halted trading at 10:58 a.m. (EST) due to an "order imbalance"—in other words, there were more sellers of Enron than buyers, and the "knife" was too sharp for the specialist to catch. Note that unlike a regulatory halt, which is market-wide, a halt for an "order imbalance" or operational failure doe not impact the ability of other markets to trade.

Which is exactly what happened . . .

While the NYSE specialist responsible for trading Enron shut down his post over the next half-hour, ECNs and other alternative venues traded over 10 million shares of Enron (NYSE Symbol: ENE), as the stock went from $2.60 to $1.10. At 11:27 a.m., the NYSE specialist resumed trading in Enron at prices discovered by these alternative markets.

The upshot: no single entity—be it exchange, specialist, or market maker—can go it alone when asked to "catch a falling knife." Efficient price discovery is the product of the entire marketplace. Provided that competing venues are informationally linked and accessible, efficient price discovery will occur at all available and open trading venues, including ECNs and alternative exchanges.[1]

V. PENNIES FROM HEAVEN

Finally, before concluding, I would be remiss in not commending Congress, the SEC, and public advocates for supporting the conversion of our equity markets to decimal pricing. In particular, the Commerce Committee was one of the critical catalysts for this positive change that, to date, has narrowed effective spreads in the most liquid stocks on Nasdaq and the NYSE by an average of 50% and 15%, respectively. This fundamental change has directly lead to enormous reductions in trading costs and put tens of millions of dollars back in the pockets of investors. Thank you for your steadfast perseverance.

VI. CONCLUSION

The dark events of September 11 continue to loom large in the collective consciousness of the world, our nation, and our financial industry. Since that day, many have called into question the wisdom of a Ptolemaic unitary model, in which our financial universe revolves around a single building at the corner of Wall and Broad Streets. The task before us is to manage risks, so that a single act of terrorism, however severe, does not endanger our system of financial markets.

Alternative markets continue to evolve toward a goal of equal standing with their more storied and traditional brethren. As Congress found in its analysis of decimalization, and as traders in Enron found in the tempest's eye of the largest corporate failure in U.S. history, it is time to set aside parochial biases. The risk of single points of failure is much too great.

I wish to thank the Subcommittee for permitting Archipelago to testify on these important matters. I would be pleased to answer your questions at the appropriate time.

---

[1] A similar example of the "Enron" phenomenon, as evidenced in Exhibit B, occurred at the NYSE on December 12, 2001, in the stock of Calpine Corporation.

EXHIBIT A

Enron Trading: 11/28/01

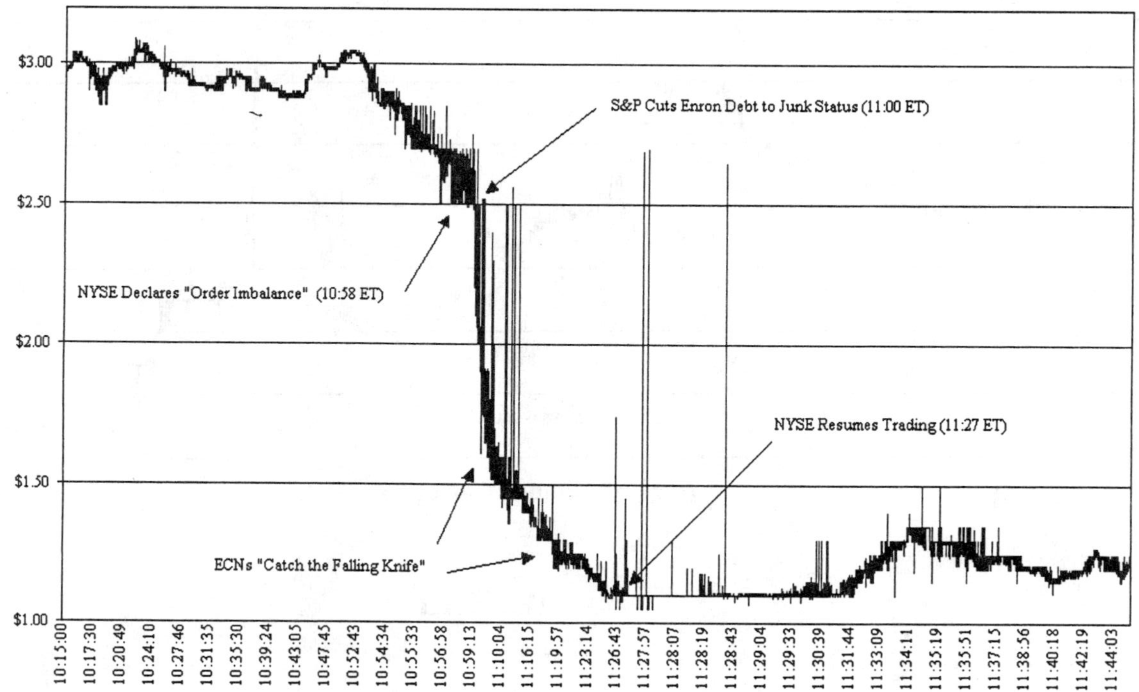

EXHIBIT B

Calpine Trading: 12/12/01

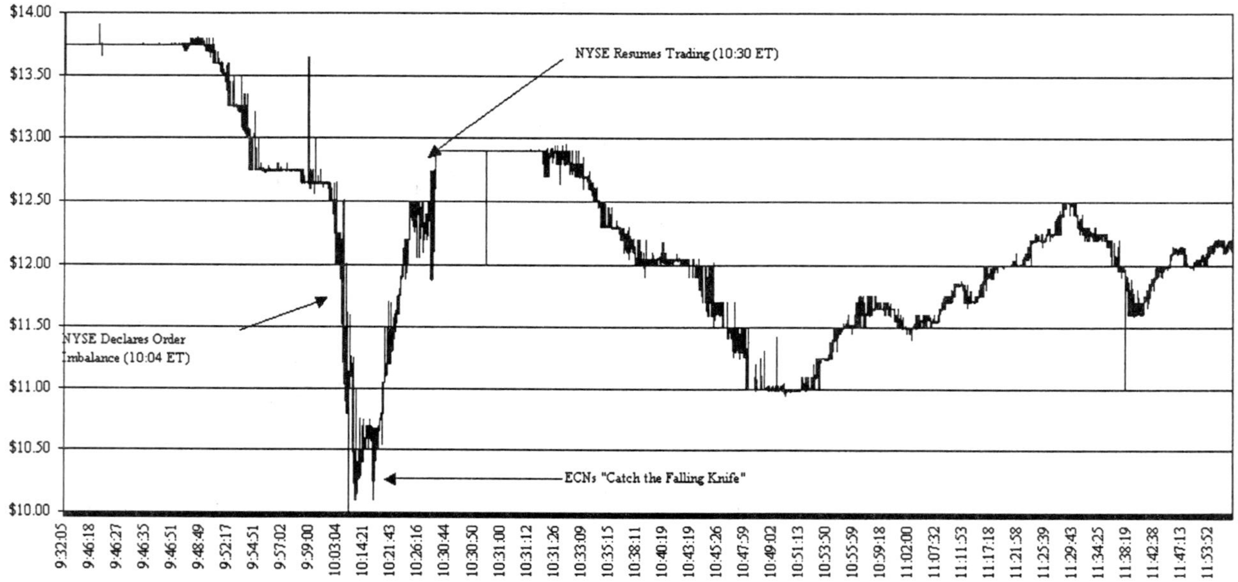

Mr. STEARNS. Thank you, Mr. O'Hara.
Mr. Randich.

STATEMENT OF STEVEN J. RANDICH

Mr. RANDICH. Thank you, Mr. Chairman, members of the subcommittee for inviting me to testify today. I welcome the opportunity to discuss Nasdaq's response to the horrendous acts of September 11 and the role of ECNs in our market.

I am Steve Randich, and I am responsible for the operations and technology of the Nasdaq stock market.

The tragic events of September 11 compelled us immediately to begin a process of evaluating the extent of any damage to Nasdaq and our market participants and determining the necessary steps to reopen the market. In doing so, we were guided by several principals. First, we would do nothing that impeded the rescue effort. Second, we would closely coordinate all of our activities with the SEC. Third, we would open our market only when major market participants were fully prepared and preferably simultaneously with the other markets. Finally, we would be as open and transparent in reaching out to assisting our members and issuing companies in a crisis as we were in everyday operations.

Because our primary backup technology centers are outside of Manhattan, our primary concern related to our ability to connect with the firms that are active in our marketplace and bring liquidity and order flow. It is critical to understand that disasters such as these are not averted by hardening any single point of failure. Rather, they are avoided by having resilience built into the network through backup connections and backup vendors. This is the key lesson from this tragedy.

In our view, the decision process to reopen the markets was a textbook example of effective cooperation among the government markets and private industry. Telecommunication, power and employee access problems created enormous complications and risks in reopening the market. In addition, there was total unanimous agreement among all participants that the equity markets should open as quickly as possible but only when we could ensure that they could operate efficiently with proper liquidity available, without additional constraints and with universal access for investors. We also believed that, given the uncertainties, it was important for investor confidence that all equity markets and their market participants begin trading simultaneously.

To achieve the successful reopening of the markets, the Nasdaq, the government and financial services industry all worked in concert.

I believe SEC Chairman Harvey Pitt said it best during his testimony before the Senate Banking Committee on September 20. "We can be justifiably proud of our market participants and the way they performed. Everyone pulled together to overcome this disaster to successfully reopen the U.S. Equities and options markets. Americans demonstrated continued confidence in our markets. With the momentum built from this experience, we will move forward to make our markets even stronger, more transparent and more vibrant."

It might be helpful to review Nasdaq's current market structure as the subcommittee looks at the role of ECNs. As the worlds' largest electronic stock market, Nasdaq is not limited to one central trading location. Rather, trading is executed through Nasdaq's sophisticated computer and telecommunication network, which transmits real-time quote and trade data to more than 2 million users in 83 countries. Last month, InfoWorld named Nasdaq 36th among the top 100 companies for information technology achievements and 5th among financial services companies.

Today, Nasdaq lists the securities of nearly 4,200 of the world's leading companies. Nasdaq's "open architecture" market structure places virtually no limit on the number of market participants that can provide liquidity on Nasdaq and places no geographical restrictions on those market participants. Indeed, Nasdaq, unlike its physical floor-based competitors, made the decision to include ECNs within its quotation and transaction systems.

At the core of Nasdaq's market structure are a group of financial firms called market-makers. More than 340 of these market-maker firms actively trade on Nasdaq, acting as liquidity providers for Nasdaq-listed securities. Also known as dealers, market-makers are unique in that they commit their own capital to Nasdaq-listed securities. Each market-maker is required at all times to maintain a bid and an offer at each of the securities in which they are registered as a market-maker.

ECNs are electronic systems that widely disseminate to third parties orders entered into the system by market-makers and permit those orders to be executed.

Mr. STEARNS. Let me have you sum up, if you could.

Mr. RANDICH. It is important to recognize ECNs, with one exception, have chosen to be brokers and, therefore, are not required to provide broad-based regulatory oversight and self-regulatory organizations.

In response to September 11, our primary focus was to ensure that our market participants were able to access our market. We believe we have the responsibility to keep up with the changing needs of the investing public to ensure that investors can buy and sell stocks quickly, efficiently and affordably, all in a fair, well-regulated market.

Mr. Chairman, we welcome this subcommittee's interest in this important topic; and I look forward to any questions that you and the other members may have.

[The prepared statement of Steven J. Randich follows:]

PREPARED STATEMENT OF STEVEN J. RANDICH, EXECUTIVE VICE PRESIDENT OF OPERATIONS & TECHNOLOGY AND CHIEF INFORMATION OFFICER, THE NASDAQ STOCK MARKET

Thank you Mr. Chairman and Members of the Subcommittee for inviting me to testify before you today. On behalf of the more than 1,200 employees of the Nasdaq Stock Market®, I welcome the opportunity to discuss Nasdaq's response to the horrendous acts of September 11, 2001, and the role of our various market participants, particularly electronic communications networks (ECNs).

I. NASDAQ RESPONSE TO EVENTS OF 9/11

The tragic events of 9/11 compelled us to immediately begin a process of evaluating the extent of any damage to Nasdaq and our market participants and determining the necessary steps to reopen the market. In doing so, we were guided by several principles: First, we would do nothing that impeded the rescue effort. Second, we would closely coordinate all our activities with the SEC. Third, we would open our market only when major market participants were fully prepared and, preferably, simultaneously with other markets. Finally, we would be as open and transparent in reaching out to and assisting our members and issuers in crisis as we are in our every day operations. A number of our participants were unable to access the network due to telephone failures. Our system could have been open on the 11th and on every day between the 11th and 17th, however we believed then, and continue to believe now, that investor protection and market integrity considerations dictated that the markets be closed. This was particularly true because of the terrible impact of the events of 9/11 on the trading facilities in lower Manhattan, including the New York Stock Exchange, the American Stock Exchange, and many Nasdaq market participants.

As to Nasdaq's technology, at no time during this disaster were Nasdaq's systems inoperative. At the time of the attacks, trading was suspended but Nasdaq's systems and network continued to operate. Because our primary and backup technology centers are outside Manhattan, our primary concern related to our ability to connect with the firms that are active in our marketplace and bring liquidity and order flow. In fact, Nasdaq continued to operate systems later than normal on Tuesday to allow firms manual access for reconciliation and mutual fund pricing and related activities. Nasdaq's systems operated virtually continuously throughout the rest of the week to allow firms to test connectivity.

Nasdaq's geographically decentralized network has several levels of redundancies, which are specifically designed to withstand these types of catastrophic events. Virtually all firms are connected to Nasdaq through a set of several Nasdaq servers on their sites and in their backup centers. Each of the servers in the Nasdaq network is connected to two distinct Nasdaq network centers using diverse telecommunications providers.

There are more than twenty Nasdaq network centers located throughout the United States—including four in the NY metropolitan area. Each of these centers is connected to both our Primary and Backup data centers. Additionally, while MCI WorldCom provides the overall management of our network, each of our critical connections is backed up by another telecommunications vendor so as to offer resiliency against a systemic provider failure.

While this may be a lengthy description, it is critical to understand that disasters such as these are not averted by hardening any single point of failure, rather they are avoided by having resilience built into the network through backup connections and backup vendors. This is a key lesson from this tragedy.[1] Therefore, one early priority was to reach out to the 344 market makers, and the 8 electronic communications networks (ECNs), that are part of the Nasdaq market. We spoke to each of these firms. We asked: Can you connect with our network? Can your employees get to their trading workstations? What problems do you foresee?

While many of our firms were not physically impacted by the disaster, many others, both market makers and ECNs, faced great challenges, in terms of personnel, technology and connectivity. Nasdaq staff worked around the clock to provide whatever support we could. This included providing alternative trading facilities, provisioning backup facilities with new equipment, testing backup and new network connections, providing assistance in acquiring emergency resources and gaining access to critical facilities in lower Manhattan.

We also reached out to the 4,190 companies that list their shares on Nasdaq. To enhance prospective liquidity, we recommended they look at buy back programs, as authorized by the SEC on an emergency basis, and get board approval if necessary.

We reached out to the SEC and other government agencies, as they reached out to us. And, we cooperated closely with each of the equity and options exchanges. The unprecedented cooperation between all market centers and with local and national governmental authorities was continuous and excellent.

I want to reiterate our appreciation of the Federal, state and local governments for their willingness to use their vast resources and regulatory powers to assist the

---

[1] See, e.g., "Key Lessons learned in attack on New York," Financial Times, December 5, 2001, "Another aspect of business continuity planning is examining how much of the organisation can be physically distributed. For example, although the headquarters of Nasdaq, the electronic stock exchange, were damaged in the September 11 attack, its distributed organisation helped it tremendously. Its two main data centers were in Connecticut and Maryland, many miles away. And its trading partners were on a network with a high degree of redundancy and only a relatively small number were unable to trade immediately after the attack. More importantly, its key people were also distributed."

markets in this time of crisis. The SEC and the City of New York were particularly instrumental in helping us open the markets as quickly and as smoothly as we did.

In our view, the decision process to reopen the markets was a textbook example of effective cooperation among the government, markets and private industry. Telecommunication, power and employee access problems created enormous complications and risks in reopening the market. In addition, there was total unanimity among all participants that the equity markets should open as quickly as possible, but only when we could ensure that they could operate efficiently with proper liquidity available, without additional constraints, and with universal access for investors. We also believed that, given the uncertainties, it was important for investor confidence that all equity markets—and their market participants—begin trading simultaneously.

After two all hands meetings, and with the strong leadership of Chairman Pitt and the full support of the SEC, Department of Treasury and Federal Reserve Board, the decision was made that trading should resume no later than Monday, September 17th. This decision was based on three primary factors. First, through the efforts of Verizon, MCI Worldcom and the affected financial firms and markets, there was a geometric improvement of telecommunications connectivity each day following 9/11. Second, the critical importance of the continuing rescue operation at the World Trade Center site made provisions for widespread physical access to financial firms and the New York Stock Exchange floor and an earlier start-up inappropriate. Third, there was complete consensus that the markets should not resume without widespread system connectivity testing that could most effectively occur over the weekend. The successful resumption of trading on Monday would be an important signal to our citizens and the world. It was accomplished by the extraordinary efforts of thousands of financial market and brokerage firm employees who collectively are owed an enormous debt of gratitude.

The SEC reassured the markets, indicated appropriate relaxation of regulatory constraints, and focused the markets on critical systems. The SEC's speedy action to ease the rules governing corporate stock repurchases was especially helpful and responsive to the needs of Nasdaq-listed companies with which we were working.

Nasdaq employees provided technological support to over 800 Nasdaq and non-Nasdaq participants including market makers, order entry firms, ECNs, other markets, and even some foreign markets seeking to re-establish their local connectivity. Many firms had to activate disaster recovery sites, which presented special technological needs.

To achieve the successful reopening of the markets, Nasdaq, the government and the financial services industry all worked in concert. The strength of the U.S. financial markets today reflects the cumulative efforts of far-sighted leadership many years ago. Of course, Congress laid the foundation with the passage and careful oversight of the U.S. securities laws.

The U.S. financial industry has demonstrated its resilience and resolve to maintain the most liquid and stable markets in the face of terrible challenges, and clearly Nasdaq's trading network has demonstrated its unique value as a part of this infrastructure.

I believe SEC Chairman Harvey Pitt said it best during his testimony before the Senate Banking Committee on September 20, 2001: "We can be justifiably proud of our market participants and the way they have performed. Everyone pulled together to overcome this disaster and successfully reopen the U.S. equities and options markets. Americans demonstrated continued confidence in our markets. With the momentum built from this experience, we will move forward to make our markets even stronger, more transparent and more vibrant."

II. NASDAQ MARKET STRUCTURE

A. Overview

It might be helpful to review Nasdaq's current market structure as the Subcommittee looks at the role of ECNs. As the world's largest electronic stock market, Nasdaq is not limited to one central trading location. Rather, trading is executed through Nasdaq's sophisticated computer and telecommunications network, which transmits real-time quote and trade data to more than 2 million users in 83 countries. Last month, InfoWorld named Nasdaq as 36th among the top 100 companies for information technology achievements, and 5th among financial services companies.

Today, Nasdaq lists the securities of nearly 4,200 of the world's leading companies, representing the entire spectrum of the U.S. economy—from information technology and telecommunications to agriculture, manufacturing and finance. Nasdaq's "open architecture" market structure places virtually no limit on the number of market participants that can provide liquidity on Nasdaq and places virtually no geographical restrictions on those market participants. Indeed, Nasdaq, unlike its physical floor-based competitors, made the decision to include ECNs within its quotation and transaction systems.

B. Nasdaq's Market Participants

At the core of Nasdaq's market structure are a group of financial firms called "market makers." More than 340 market making firms actively trade on Nasdaq, acting as liquidity providers for Nasdaq-listed securities. Also known as "dealers," market makers are unique in that they commit their own capital to Nasdaq-listed securities. Each market maker is required at all times to maintain a bid and an offer in each of the securities in which it is registered as a market maker. By being willing to buy and sell stock—using their own funds—market makers add liquidity to Nasdaq's market, ensure that there are always buyers and sellers for Nasdaq-listed securities, and enable invertors' trades to be filled quickly and efficiently. Market makers adhere to strict trading regulations and are required to:

- Disclose their buy and sell interest by displaying continuous two-sided quotes in all stocks in which they choose to make a market.
- Display customer orders in their quotes in Nasdaq or in the quotes of ECNs, in compliance with SEC Order Handling Rules.
- Honor their quoted prices and report trading in a timely manner. Failure to do so can lead to disciplinary action.

In addition to market makers, the Nasdaq network also connects alternative trading systems into the market, such as ECNs. ECNs are electronic systems that widely disseminate to third parties orders entered into the system by market makers and permit those orders to be executed against. Preliminarily, it is important to recognize that ECNs, with one exception, have chosen to be brokers and, therefore, are not required to provide broad based regulatory oversight as are self-regulatory organizations.

The largest ECNs are: (1) Instinet, which is majority owned by the British firm Reuters, and (2) Island. Other ECNs include Bloomberg's Tradebook, Archipelago (which recently merged with the Pacific Stock Exchange), and Redibook (which recently agreed to merge with Archipelago). With the exception of Archipelago, which will operate in part as an affiliate of a regulated exchange, the ECNs are regulated just like other broker-dealers.

These ECNs provide electronic facilities that investors can use to trade directly with each other. Additionally, they provide investors with an anonymous way to enter orders into the marketplace. ECNs operate as order-matching mechanisms and do not maintain inventories of their own or risk their own capital. ECNs are not required to maintain continuous two-sided quotations in the securities that they trade.

Nasdaq recognizes the unique role that ECNs play as part of an integrated Nasdaq Stock Market. In 1997, the SEC required ECNs to allow access to their systems by non-subscribers. As a result, ECNs are integrated into the National Market System and investors have benefited through enhanced liquidity.

III. CONCLUSION

In response to 9/11, our primary focus was to ensure that our major market participants were able to access our market; each of them has an important role to play. At Nasdaq, we believe that we have a responsibility to keep up with the changing needs of the investing public to ensure that investors can buy and sell stocks quickly, efficiently, and affordably, all in a fair, well-regulated market.

As we move forward, all of Nasdaq's efforts to improve its market structure, including SuperSOES (small order execution system) and SuperMontage, will impact the quality and depth of information that we can provide to investors. Today, investors in Nasdaq securities can only see the aggregate trading interest at the best prices to buy and sell. When implemented next year, SuperMontage will display the total amount of trading interest in Nasdaq at the best bid price and at the best offer price, as well as two trading increments away from those prices. This expanded display will increase transparency by allowing customers and other market participants to see greater depth of market. As a result, investors will have more information on which to make better-informed trading decisions.

Nasdaq's open architecture market structure fosters innovation in the creation of new products and services, new market participants—such as ECNs—and new business models for the ultimate benefit of investors.

Mr. Chairman, we welcome the Subcommittee's interest in these important issues, and I look forward to any questions you and the other Members may have.

Mr. STEARNS. I thank the gentleman.

Mr. Jamaitis, we welcome your opening statement. Roughly about 5 minutes.

STATEMENT OF KEITH R. JAMAITIS

Mr. JAMAITIS. Good morning, Chairman Stearns, members of the subcommittee. I am Keith Jamaitis, Chief Operating Officer of NYFIX. I appreciate the opportunity to testify on behalf of NYFIX Millennium regarding the important role that electronic communication networks, ECNs, and alternative trading systems, ATSs, played in the wake of September 11.

It is our view that the SEC's vision of fostering competition among execution centers such as ECNs, ATSs and exchanges contributed directly to the swift recovery of our markets after the September 11 attacks. We are pleased to offer the following remarks which will touch upon NYFIX's activities and role in helping the equity markets reopen following the tragic events of September 11.

NYFIX Millennium is a registered NASD broker-dealer that operates an ATS. The NYFIX Millennium ATS is an electronic execution venue focusing specifically on exchange-listed securities. The company was founded in order to provide a superior execution platform by leveraging the technological infrastructure and order routing volume of the NYFIX network. Our mission is to provide high-quality executions in the listing trading arena through anonymous and efficient matching of pass-through and conditional order flow.

The NYFIX financial technology infrastructure delivers approximately 500 to 600 million shares of listed trading volume to the New York Stock exchange each day. This order flow represents a substantial portion of the block trading volume delivered to the New York Stock Exchange electronically and executed in the crowd.

NYFIX Millennium was a victim of the September 11 attacks along with the rest of the financial services community and the rest of the world. Our sympathies are with our professional peers and all the people and families affected by these events.

During and after the World Trade Center attacks, the NYFIX service bureau staff kept focused on helping our customers and industry recover from these disastrous events. NYFIX is uniquely positioned with offices in Stamford, Connecticut, as well as New York City. The corporate headquarters and the help desk are located in Connecticut. The help desk, while normally servicing trading systems and issues with our customers, quickly became an industry information hub in the wake of attacks.

NYFIX maintains many critical communication links that cross-connect the electronic trading systems that service our equity markets. These connections are very comprehensive and allow trade data to flow from buy-site institutions to sell-site institutions to execution destinations such as the New York Stock Exchange, third party market destinations, and ECNs and ATSs.

NYFIX Millennium's connectivity with the financial services industry allowed us to share information among the financial institutions. Most importantly it gave our operations staff a clear assessment of the technological impact of the World Trade Center disaster. NYFIX immediately undertook an inventory of all internal systems that were affected. The data centers have dual locations in

Carlstadt and North Bergen, New Jersey, so the effect to core systems was minimal. The major problem was telecommunication outages. The NYFIX network lost over 60 percent of its telecommunications data services into New York City. Immediately after the disaster, our data network into New York Stock Exchange was reduced by approximately 80 percent.

System recovery operations began immediately from the Stamford, Connecticut, office. Our telecommunications carriers were notified, and specialized task teams were deployed to resolve system issues. Once in-house issues were under control, the help desk and account management staff attempted to contact each of our over 100 broker-dealers in our network.

The primary problem remained telecommunications data circuits. As a true business partner, we began to address the secondary issues of relocating customers who have been displaced from the World Trade Center and other downtown locations. The specific steps taken included shipping multiple systems to newly established customer disaster recovery sites, assembling over 100 additional systems into inventory, deploying our application engineer staff to alternate sites to work on network configurations and new trading system installations, reconnecting several customer systems via the Internet, providing Internet access through the NYFIX Millennium network out to our customers, making our office space in New York available for temporarily displaced customer users, reengineering some trades clearing processes that our customers use on the New York Stock Exchange.

It is important to note that our service bureau is a significant investment. It is a dual data center model with multiple redundant telecommunications circuits that allow for maximum reliability and flexibility. Our commitment to providing the highest level of data services made the systems recovery effort possible. It is difficult to measure the value of the substantial investment in multiple backup systems and facilities. The best technology is good only to the extent that it is available.

We understand our role in the industry and the responsibility associated with it. We understand our responsibility to our over 1,700 users on institutional trading desks. We understand our responsibility to the over 180 broker-dealer clients on the floor of the New York Stock Exchange. We fulfill these responsibilities on a daily basis.

We will continue to invest in reenforcing our technology platform to deliver the best service possible to the investing public. NYFIX Millennium's disaster relief procedures and systems performed well under the extremely adverse conditions of September 11. We are consulting with the financial services industry on adopting and implementing similar best practices, including dual mirrored data centers on dual power grids located at different geographic locations; adopting formal procedures to require regular system recovery testing; and adopting NYFIX disaster recovery policies and procedures to be applied to our clients' proprietary systems.

The events of September 11 demonstrated the value and strength of our technology. Our NYFIX network remained online during the terrorist attacks and were 100 percent prepared for markets reopening. In fact, we experienced our greatest share volume on September 17, routing approximately 1.2 billion shares to the floor of the New York Stock Exchange through our systems.

In conclusion, upon the approval and coordination of the regulatory community, our equity markets were prepared to open on September 17. We are proud to have been a significant contributor to their successful opening and operations on that day. Our efforts have continued through to the present to improve and refine our disaster recovery capabilities. We are particularly pleased with our service to the investing public and our clients as coordinated through the New York Stock Exchange, its member firms, and other industry leaders participating on this panel today. Thank you.

[The prepared statement of Keith R. Jamaitis follows:]

PREPARED STATEMENT OF KEITH R. JAMAITIS, SENIOR VICE PRESIDENT, CHIEF OPERATING OFFICER, NYFIX, INC.

Chairman Stearns, members of the Subcommittee: I appreciate the opportunity to testify on behalf of NYFIX Millennium, Inc. ("NYFIX Millennium") regarding the important role that electronic communication networks ("ECNs") and alternative trading systems ("ATSs"), played in the wake of September 11.

It is our view that the SEC's vision of fostering competition among execution centers such as ECNs, ATSs and exchanges contributed directly to the swift recovery of our markets after the September 11 attacks. We are pleased to offer the following remarks, which will touch upon NYFIX's activities and its role in helping the equity markets reopen following the tragic events of September 11.

I. ABOUT NYFIX MILLENNIUM

NYFIX Millennium is a registered NASD broker-dealer that operates an ATS. The NYFIX Millennium ATS is an electronic execution venue focusing specifically on exchange-listed securities. The Company was founded in order to provide a superior execution platform by leveraging the technological infrastructure and order routing volume of the NYFIX Network. Our mission is to provide high quality executions in the listed trading arena through anonymous and efficient matching of pass-through and conditional order flow.

The NYFIX financial technology infrastructure delivers approximately 500-600 million shares of listed trading volume to the New York Stock Exchange each day. This order flow represents a substantial portion of the block trading volume delivered to New York Stock Exchange electronically, and executed in the crowd.

II. SEPTEMBER 11

NYFIX Millennium was a victim of the September 11 attacks along with the rest of the financial services community, and the rest of the world. Our sympathies are with our professional peers and all the people and families affected by these events.

During and after the World Trade Center attacks the NYFIX service bureau staff kept focused on helping our customers and the industry recover from these disastrous events.

NYFIX is uniquely positioned with offices in Stamford, Connecticut as well as New York City. The corporate headquarters and the Help Desk are located in Connecticut. The Help Desk, while normally servicing trading and systems issues for our customers, quickly became an industry information hub in the wake of the attacks.

NYFIX maintains many critical communications links that cross-connect the electronic trading systems that service our equities markets. These connections are very comprehensive and allow trade data to flow from buy-side institutions to sell-side institutions to execution destinations such as the New York Stock Exchange, third market destinations, ECNs and ATSs.

NYFIX Millennium's connectivity with the financial services industry allowed us to share information among the financial institutions. Most importantly, it gave our operations staff a clear assessment of the technological impact of the World Trade Center disaster.

NYFIX immediately undertook an inventory of all internal systems that were affected. The data centers have dual locations in Carlstadt and North Bergen, New Jersey, so the effect to core systems was minimal. The major problem was telecommunication outages. The NYFIX network lost over 60% of its telecommunication data services into New York City. Immediately after the disaster, our data network into the New York Stock Exchange was reduced by approximately 80%.

System recovery operations began immediately from the Stamford, Connecticut office. All telecommunications carriers were notified and specialized task teams were deployed to resolve systems issues. Once in-house issues were under control, the Help Desk and account management staff attempted to contact each of the over 100 broker-dealers in our network.

The primary problem remained telecommunication data circuits. As a true business partner, we began to address the secondary issues of relocating customers who were displaced from the World Trade Center and other downtown locations. Specific steps taken included:

- Shipping multiple systems to newly established customer disaster recovery sites for immediate use;
- Assembling over 100 additional trading systems into inventory;
- Deploying application engineers to alternate customer sites to work on network configuration and new trading systems installations;
- Reconnecting several customer trading systems via the Internet;
- Providing Internet access through the Millennium network to our customers;
- Making office space available in both of our New York office locations to temporarily house displaced users; and
- Reengineering several trades clearing processes associated with our New York Stock Exchange customers.

It is important to note that our service bureau is a significant investment. It is a dual data center model with multiple redundant telecommunications circuits that allow for maximum reliability and flexibility. Our commitment to providing the highest level of data services made the systems recovery effort possible. It is difficult to measure the value of the substantial investment in multiple backup systems and facilities. The best technology is good only to the extent that it is available and reliable. We understand our role in the industry and the responsibility associated with it. We understand our responsibility to our 1700 users on institutional trading desks. We understand our responsibility to the over 180 broker-dealer clients on the floor of the New York Stock Exchange. We fulfill these responsibilities on a daily basis.

We will continue to invest in reinforcing our technology platform to deliver the best service possible to the investing public.

NYFIX Millennium's disaster relief procedures and systems performed well under extremely adverse conditions. We are consulting with the financial services industry on adopting and implementing similar best practices, including:

- Dual mirrored data centers with dual power grids, located at different geographic sites;
- Maintaining multi-carrier WAN infrastructure of data circuits;
- Adopting formal procedures requiring regular systems recovery testing; and
- Adopting NYFIX disaster relief policies and procedures to be applied to our clients' proprietary systems.

The events of September 11 demonstrated the value and strength of our technology. Our NYFIX network remained online during the terrorist attacks and were 100% prepared for the market's reopening. In fact, we experienced our greatest share volume on September 17, routing approximately 1.2 billion shares to the floor of the New York Stock Exchange through our systems.

III. CONCLUSION

Upon the approval and coordination of the regulatory community, our equity markets were prepared to open on September 17. We are proud to have been a significant contributor to their successful opening and operation on that day. Our efforts have continued through the present, to improve and refine our disaster relief capabilities. We are particularly pleased with our service to the investing public and our clients as we coordinated with the New York Stock Exchange, its member firms and the other industry leaders participating on this panel in ensuring that our equity markets got back to business.

Mr. STEARNS. I thank the panel very much. Let me first of all say congratulations on how quick that you did get back to operation. And it shows the ingenuity and the entrepreneurship for all of you to get back so quickly. Mr. Shimkus and I, as he mentioned, were able to go up to Ground Zero shortly after September 11, and we are so impressed with Verizon, how quickly not only they got the cell phones back, but the land lease lines, considering the flooding and all the damage there.

The first thing I wanted to speak to is the ECNs. Lots of them are located in New York City, and, I mean, is there a reason why you would have to be located in New York City considering that?

Mr. TOWNS. I will answer that, Mr. Chairman.

Mr. STEARNS. Mr. Towns is a little sensitive about that.

Mr. ANDRESEN. Certainly in any business there is an advantage to geographic proximity to your customers from a relationship standpoint and everything else, but there is certainly no direct need from a technological standpoint to have our one and only business in New York. Island has a data center at 50 Broad Street right down from our distinguished colleagues at the New York Stock Exchange. We also have a site out in Secaucus, New Jersey.

Mr. STEARNS. It looks like your—some of them are located at Times Square and also Broad Street. Mr. Jamaitis has mentioned that he has something out in Stamford, Connecticut.

Mr. JAMAITIS. Our headquarters and operations are based out of Stamford, Connecticut.

Mr. STEARNS. So that if we had another calamitous event in New York City down near the World Trade Center, the question is, is there enough redundancy that the ECNs would be affected or not?

Mr. JAMAITIS. The policy of our network and our systems, we have our core data centers located in two locations in New Jersey, both Carlstadt and North Bergen; then that tertiary fail over capability on Wall Street, and we can also fail over back to our Stamford offices. So we feel being geographically remote and diverse has been a great advantage, especially weathering, you know, the disasters we saw.

Mr. STEARNS. Mr. Andresen, you mentioned that—I think you said you do 460 million shares a day.

Mr. ANDRESEN. That is correct.

Mr. STEARNS. How many employees do you have?

Mr. ANDRESEN. We have 138 employees.

Mr. STEARNS. Okay. Am I free to ask what your revenue is a year? Not the revenue of the trade—you don't have to give this, but if you feel comfortable.

Mr. ANDRESEN. We are a private company, so our investors will come find me after the meeting if I disclose too much.

Mr. STEARNS. Mr. Steinmetz, you said in your testimony that the bond markets and currency markets suffered greater devastation on September 11 than the stock market, and I don't think most people realize that fact, but were able to resume trading more quickly. Can you perhaps give us more detail on those markets and why they were able to resume trading quicker, more quickly?

Mr. STEINMETZ. The bond markets and the currency markets have very different market structure particularly as it relates to the networking versus the standard mainframe communications, if you will. There are approximately 49, I believe, trading platforms and fixed income that people trade on. There are numerous amounts of instruments, a lot more instruments than equity instruments, and they trade all over through different networks. So, therefore, even though we had such terrible devastation to some of the bigger companies that trade in those markets, they were able to get up and running on alternative systems.

Mr. STEARNS. Okay. Mr. Bang, you had mentioned that—you used the word "government-sponsored monopoly," and you talked about decentralizing versus centralizing. Yet as I understand, the New York Stock Exchange has allowed ECNs to access the intermarket trading system through NASDAQ. Why is this access insufficient for ECNs, and why would this not show that, you know, there is this accessibility?

Mr. BANG. Mr. Chairman, the access into NASDAQ's intermarket trading is a partial solution. It would give us the ability to display our quotes in the consolidated quotes system for listed securities. However, we would also be subject to the ITS rules that Matt Andresen pointed out and described as somewhat incumbent or problematic for extending the services out to our clients with the features that they are accustomed to, which is one of immediacy, which is one of the ability to access liquidity where liquidity is to be had. And the problem of being subject to trade through rules and such, where you have to go outside and access liquidity over the ITS systems against other market centers, that can take up to a minute to respond to those outbound orders.

Mr. STEARNS. Mr. O'Hara shows this graph of Enron trading and showing how ECNs stepped up to the plate. And then actually when the markets came on board, they used as a reference plane the actual trading values of Enron to establish a base from which to work. So, would the market take care of itself? Are you saying to us, Mr. Bang, that you expect to let the market eventually decentralize this, or are you looking toward some kind of outside influence, either policy from the government or policy de facto, but never regulated by the SEC or somebody, to provide what you have indicated is a monopoly—to open it up to more competition, which in the end would mean if we had a calamitous September 11 again, there would be more redundancy?

Mr. BANG. Yes, I believe that is an accurate statement. You know, just as the rule 390 was done away with, it gave—opened up a certain element of competition for printing the New York Stock Exchange list of securities off other exchanges, and that was a good development, likewise I would say if ECNs have the ability to represent their investors' and clients' interest in the national, quote montage, that additional transparency would create, you know, liquidity to the national market system and would provide alternative venues to trade away from, which Ms. Kinney talked about, a central hub.

Currently we have, you know, very high centralization for the New York Stock Exchange list of securities on the exchange, that central hub. I think it would behoove the industry to have more than one hub or alternative hubs with subsequent spokes into them. And ECN certainly is a venue that is well suited to fulfill that.

Mr. STEARNS. I think my time has expired.

The gentleman from New York.

Mr. TOWNS. Thank you, Mr. Chairman.

Ms. Kinney, would you please respond to the points that Mr. Bang made about access to the New York Stock Exchange system?

Ms. KINNEY. Well, I think he pointed out that he has access to our systems through the NASDAQ intermarket system, but he also pointed out that the ECNs are unwilling to conform to one of the principles of ITS, which is price protection for customers' orders, and as such, they have been precluded from participating.

We think that price protection is one of the foundations and hallmarks of the national market system, that a customer is entitled to the best price wherever it exists, and if an order arrives in a certain venue, any one of these, including our own, and there is a better price elsewhere, the customer should be protected at that price.

Mr. O'HARA. If I may just clarify the record, Archipelago does today—is the only ECN to access the New York Stock Exchange through our friends here at NASDAQ through ITSKs. We have a different customer base than others, but they have told us that they want access through ITSKs, and we access New York through ITSK as well as their proprietary DOT line as well. So the fact is that our customers are happy with it, and we give them that service.

Mr. ANDRESEN. I just wanted to thank you, Congressman Towns. From Island's perspective our customers are, as Mr. O'Hara pointed out, a different group of customers who have a different value proposition that they value.

On point—in addition to your point, Mr. Chairman, there is a difference between what people see on ECNs and what people see in the national quote. If it is—we are trading in a NASDAQ stock, say Intel, and Island has the best price, if someone goes to NASDAQ and says, hey, NASDAQ, where is the best price in Intel, they will see Island's quotes. However, if the best price is queried for QQQ or for IBM, ECNs right now are not permitted to show their quotes in the national quotes.

Now, Island distributes all of its market data for free over the Internet to make sure investors still have access to this information, but there is a difference between what is actually happening and what has received the stamp of the national market system.

Island does have issues with the intermarket trading system. Specifically our view is that our marketplace and liquidity we have built are significantly new traders doing new trades, and they rely on the speed and reliability and cost that Island gives. I think a great analogy to ITS is I used to be—Mr. Towns, you will be pleased to know that I did move back to New York from New Jersey. I couldn't stay away. But I used to drive through the Holland Tunnel every morning. Every morning I get up there, at first I had to go to the full-service lane because I didn't have EZ Pass, so I had to get my 4 bucks out and have to pay. It took forever. So I got an EZ Pass, and I was able to shoot right through. EZ Pass was worth a lot to me because I was able to get to work like a half an hour earlier.

What ITS does is basically says, well, Island, if you would like to open up an EZ Pass booth here and transact electronically, that is great, but you can't allow anyone to go through your gate until the guy on the other end of the line, the full-service lines, breaks his $100 bill. You have sort of allowed competition in allowing us to transact electronically, but you slowed Island down to being no faster than the slowest other toll lane. And I think as Wick Simmons pointed out in his letter to Harvey Pitt, these trades will not move back somewhere else if Island has to slow down. They will simply disappear or move offshore.

Mr. BANG. If I may add, since the advent of decimalization——

Mr. TOWNS. Go ahead. I want to come back to Ms. Kinney. Then I will go to you.

Ms. KINNEY. I think that Matt makes a point about his customer base. I think that Matt's and Island's sole value proposition is speed, and the markets that are linked have a responsibility to provide price protection to their customers. The fact that the value proposition is exclusively speed, his customers enter that market knowing what the value proposition is, pursue it, and receive it. And it is not conceivable to me to understand how in—and he cites in his own testimony about the QQQ, they have a significant market share there. The value proposition has obviously worked in that model. And we continue to compete with them in that product, but we are offering the best price at the lowest spread with price protection.

So, again, it is going to be customers choosing among the markets, but there have to be fundamental and underlying value propositions for each of us, and customers are choosing those very freely today in the market, whether it is in Enron or in the QQQ.

So I think the markets are very accessible at the moment. Information is available, and customers are taking advantage of the models they were pursuing.

Mr. TOWNS. I thought I would get a little extra time being I am from New York. Thank you, Mr. Chairman.

Mr. BANG. It is exactly that choice that we are advocates of. And we believe that it would behoove industry to provide that choice, the true choice for investors, and the way to do that is to give—I would say, one way to give it is to provide ECNs direct access into national quotes. The trade-through issues are much less of an issue today as we are in a decimalized environment than it was prior to decimalization. We were describing different increments. A penny here and penny there is a small price to give up for the certainty of accessing a certain pool of liquidity or a given price,and we believe that investors should have that choice and be able to make that decision.

Mr. O'HARA. And just to confirm, I think, what you are hearing here is a marketplace right here, is that different customer bases want different services, and that through competition companies should be allowed to provide it. And I think from a congressional standpoint or from a regulator standpoint at the SEC is as long as policies are established that allow competition to grow within a certain structure, that you are exactly hearing the marketplace here: Different customers want different things. Some want ITS access, some don't.

I think if we allow things to evolve as they currently are, we will find the solutions that I think we are looking for at this hearing today.

Mr. TOWNS. Thank you, Mr. Chairman.

Mr. STEARNS. Sure.

I would say we probably have a second round here. Mr. Shimkus.

Mr. SHIMKUS. Thank you, Mr. Chairman.

I want to kind of go back and focus on the two hats that I wear with respect to my telecommunications hat and, of course, as the consumer protection issue. I was able to during the last trip get some pictures. I know you all can't see them, but you all will recognize some of them. This is 140 West and building 7, about three stories high of rubble up against there. I will share these with my colleagues. This is 140 West, rubble almost four stories high. These are the windows that got impaled—you can see that is probably a 20-foot by 20-foot hole in 140 West—by the beams that shot across.

This is the one I really want—there is a couple that I really want to—I didn't have all of them, but this is one of the switching rooms in 140 West. If you can look closely, here they are, the cabinets one of the switching rooms. Now, this is of the basement of 140 West, all these cables. And with the chairman, we walked down there, and it was—they are still trying to dry it out with fans down there. You can see the multitude of cables, such that what they had to do was they brought the cables outside the building seven stories high into a window. That is the effort that was made. Here are some good workers hand-twining the little phone lines, millions.

My question is—the question that I am posing is—goes back to a debate that we have here in Washington on the telecom side, which is regional Bell operating companies versus competitive local exchanges. Fun, huh? This whole facility, 140 West, is operated by a regional Bell operating company, an RBOC as we call it in the vernacular, versus a CLEC which has the ability to be there. The difference is this 20-some-odd-story building with all that infrastructure versus a CLEC being one or two of these cabinets.

In the event of another major catastrophe, who do you think is better able to respond in a timely manner, an RBOC with vested interest in infrastructure or the competitive local exchange? And how do we look beyond there to make sure that we have the capacity to meet major infrastructure needs? And I will just throw it open to those who want to dare walk into this mire.

Mr. O'HARA. Well, I certainly don't pretend to be an expert on RBOCs and the CLECs, although I have obviously read some things about it. I think what you can take from our industry to that is as long as there is open, free competition, you know, and a level playing field, I think you will find, you know, through literally thousands of years of history now that you will get to the place that you want to be, and that is the most efficient, you know, price, best service, that type of thing.

You know, in our industry that is where our Archipelago and, I know, some of my colleagues come from. As long as there is a fair regulatory and level playing field, let's have it, and in the end consumers and investors will win. I would assume, given the hundreds of thousands of years of history of competition since the history of capitalism, as to your specific question, that is where you would end up if you had that type of a level playing field.

Mr. STEINMETZ. I think what your question really underscores is an interesting connectional bridge between some of the market structure issues that were discussed here today as well as the technological issues that we know existed. And if we step back and sort of bridge those two and combine them, we actually will be able to determine that the best way to do it is to have a broader networking communications area as opposed to one centralized area.

Now, whether those things include exchange floors or whether they include ECNs, it is possible to include all of them in a general market structure. But if we can, going forward, not necessarily rely on either one of those solutions, but rather rely on a joint solution of multiple players that can interconnect better, then we would have the backup systems and redundancy and contingency plans necessary to exist in any other catastrophe that might come our way.

Mr. RANDICH. I think the lesson learned in those photographs is you can't put all your eggs in one basket. We as enterprises need to take the opportunity to ensure that we have diversity, diverse connections to our customers, and the only way we can do that is if we have a number of valid choices, and those choices range between the local access carriers, the long-haul carriers, as well as the RBOCs. NASDAQ uses all three of those types of carriers to provide that diversity nationwide.

Mr. SHIMKUS. Do you think it also makes a statement—and I have no real agenda here, but I am just trying to debate for your industry—you need to have secure lines of communication, and in essence our eggs are in one basket. If there is diversification, doesn't it argue that CLECs ought to be a little more infrastructure-independent instead of reliant?

Mr. O'HARA. I think you have us puzzled there.

Mr. SHIMKUS. Infrastructure-independent means one cabinet versus a 20-story building.

Mr. O'HARA. I would underscore what Mr. Randich said, that you can't have all your eggs in one basket as long as you have a level playing field so they can compete, and not knowing what the CLECs—you know, what their long regulatory structure looks like, as long as they have the ability to raise capital, if they have a good idea and build that infrastructure, and there are no impediments, I think it will take care of itself.

Mr. BANG. I would like to add to that we learned that our dependency on the West Street telecommunication hub for all of us was clearly too large, and for our customers. And I think that each one of us have, you know, taken measures to decrease that dependency, and so have our customers. But we probably didn't realize quite how large our dependency was in that particular area, and that is something that we have to be cognizant of.

Mr. STEARNS. The gentleman's time has expired.

Mr. SHIMKUS. If I could say this is one of the first times that I have been able to throw out acronyms that may have befuddled the panel, so this is a big victory for me.

Just to close, I want to welcome Mr. O'Hara, who is from Chicago, Illinois, a great State, and also appreciate your work in the Eastern bloc countries, Lithuania, and trying to work on their entry into NATO. I know you spent time on that effort.

Mr. STEARNS. Mr. Deal, gentleman from Georgia.

Mr. DEAL. Thank you, Mr. Chairman.

During the discussion of this, the ECNs and the Internet have been compared, and as you may know, this committee is constantly wrestling with the issue of data integrity and security issues as it relates to the Internet. And I would like to ask the ECNs if there is a concern or an issue with regard to data integrity within your systems, and if so, what are you doing to allay those concerns?

Mr. ANDRESEN. At Island we have the advantage in our business model and the business model of most of the people up here, unlike the Internet which is eventually giving access to everybody, we are giving access to a select number of professional brokerage firms. Island has about 700 different brokerage firms connected to us. The way they connect to us is either through a frame relay from MCI or AT&T or through a direct point-to-point line. Because of this, because it is a small universe of actual users, although an immense amount of data, it is much easier to control our problems than it would be for, you know, eBay or Amazon to control theirs. So that is an innate advantage to our business model.

Our biggest concern on data integrity is something that Mr. Steinmetz from Instinet brought up earlier, and that is the importance of anonymity. The most important thing that we can do for the integrity of our marketplace is information about who bought what where is not available in an asymmetric fashion, so Island gives out all of its market data for free over the Internet in real time; every piece of information, that is, except who it is.

I remember when I was trading, and any time Goldman Sachs made a low offer on my stock, I got a terrible feeling in the pit of my stomach. If I saw a low offer on Instinet, I was—I wondered whether it was Goldman, but I could never know. That is better for Goldman, that is better for me, and that is better if it happens to be my mom trading the stocks as well.

Mr. DEAL. Anyone else?

Mr. STEINMETZ. We have dealt with the security issues on several levels. The first is a simple encryption level on the base level making sure that it is secure, as well as certifications on the terminal-by-terminal level to assure that, again, there is further security depending upon what customer connects where. So certain customers would like certain levels of security, and they can settle with the encryption on the base level. Others require something a little further and therefore get the certification on the higher level.

In either case, though, the Internet technology has not caused a slowdown in the connectivity, which is crucial. As has been mentioned on this panel already, speed of customers interacting with the market is essential toward better execution.

With that in mind, as long as we can assure the security, and the Internet allows that, and still provide the speed, then we shouldn't have much of the problems in the security area using the Internet.

Mr. DEAL. Mr. O'Hara, you may be from Chicago, but thanks to Margaret Mitchell, that is a good Georgia name as well.

I would like to ask you, you say that the linkages between the electronic facilities could have allowed trading to occur after September 11 at an earlier timeframe. You posed the question as to whether the industry is prepared to move to a decentralized model. I would like to ask you if you would elaborate on what is preventing the industry from moving to a decentralized model, and what are the disadvantages of moving to a decentralized model, if any?

Mr. O'HARA. I know our friends at the New York Stock Exchange would differ with us as to the latter part of your question, but as to the former, I think we certainly have over the last 10 years, especially the last 5 years, moved rather quickly toward decentralization in large part. We have a lot of work to do yet and a lot of wood to cut, but in large part, because of some of the people here and others that aren't here, NASDAQ, through the order handling rules, the SEC opening up NASDAQ to ECNs, that in large part—creating many networks within NASDAQ, that has allowed us to decentralize in part.

Also, just technology is at a place today where—for instance, where Island or REDIBook and Bloomberg—for instance, we have proprietary lines that we put in between each other, not government-mandated, the SEC hasn't told us to do that, but we have on our own said our customers want this, so we have done that. In the current environment, given the regulatory environment allowing this to happen, and second, given that technology is where it is today—this couldn't have happened most likely 15, 20 years ago, especially predevolution of AT&T—because technology is where it is today, it has allowed decentralization to occur.

Now, you ask the question on the down side there are some who argue, and I am not sure if Ms. Kinney is going to argue this point, but that all orders should be brought to one location. So, in other words, the deepest pool of liquidity is where people will get best price or price improvement or that type of thing. The SEC floated that idea with a central limit order back with NASDAQ, and I think everyone from alpha to omega said that is not a good idea, that you don't want to bring all your orders to one place in part because of what we learned from September 11.

What we can do and what actually exists in part today is to have a network of virtual—a virtual world where everyone is talking to each other, we are talking to Island, we are talking to the New York Stock Exchange, and customers can access prices at different places as we compete against each other, although we are all connected to each other.

So the upshot is we are working toward it. I think we are partially there, but we have some wood to cut toward getting there.

Mr. DEAL. I assume you are saying since you are virtually unregulated by the government, you don't need our help in moving that direction.

Mr. O'HARA. We are certainly—I see my friends at the SEC over there—we are very much regulated by the government, and we should be, quite frankly. But the fact is, and I think is a real credit to the SEC, that over the last 5 years they have allowed a lot of competition, they have taken some chances, rational chances, and I think in the end people are seeing results and cheaper, more efficient services for customers and investors.

Mr. STEARNS. The gentleman's time has expired.

Mr. DEAL. Could I ask unanimous consent for Ms. Kinney to respond to the question?

Mr. STEARNS. Sure.

Ms. KINNEY. The market models today and the New York Stock Exchange has a lot of competition. We certainly open every morning and have every regional exchange, NASDAQ, all the ECNs competing with us very aggressively, as you can hear from all of these testimonies. ECNs represent only 3 percent of the trading volume on the New York Stock Exchange, and you contrast that to the kind of activity that you see in the NASDAQ model where ECNs may be as much as 30 to 40 percent of the activity in Microsoft, for example.

So the marketplace and the structures exist for the competition to occur. I would like to think that the New York Stock Exchange innovates very aggressively to compete. We have been successful there. We do believe that some centralization of the orders flow does provide the best prices.

That said, we have to constantly be alert to those that are appearing today so that we are providing the kinds of services that customers want. At the end of the day, if they don't want what we are providing, they are going to send the order to Matt Andresen or to Mr. O'Hara or to anybody here and get an execution from them.

So the New York Stock Exchange will compete. I think I am happy to report that so far, so good. But we are very alert. These are competitors we have a lot of respect for.

Mr. DEAL. Thank you, Mr. Chairman.

Mr. STEARNS. The gentleman from Arizona.

Mr. SHADEGG. Thank you, Mr. Chairman. I commend you for holding this hearing. I appreciate the unanimous consent to file my opening statement since I was a little late.

Mr. STEARNS. So ordered.

Mr. SHADEGG. Many of you have focused on the benefits of an Internet packet switched network system over a traditional circuit switch system. I guess I would like to kind of go the next step and ask you if you would comment on or elucidate us on what kind of challenges you think we need to think about to such a system given the possible threat of a cyberterrorist attack as opposed to a physical attack like the one we experienced on September 11.

Ms. KINNEY. I don't know that we would be supporting necessarily an Internet kind of connectivity. I think that Matt made the point, which we would agree completely with, and that is we are fortunate in that we can invite into the marketplace our members, those that have qualified and have certain requirements for capital, knowing their customers, and a variety of other regulatory requirements that at least make them on some level known to us.

The New York Stock Exchange operates a private network among its members and therefore is protected in many ways from the security issues Mr. Deal questioned. And, in fact, following September 11, a number of the member firms have asked the Exchange, the Securities Industry Automation Corporation and, I think, others on this panel to be part of that discussion, to have for our industry a very private network that could be insulated in some way from some of the things that we experienced on September 11.

I think many of our customers, many member firms, were mistaken when they thought if they bought from two carriers a service, that would be protecting them, one as a primary, one as a backup. They sadly learned that some of those were running through the exact same channels or cable.

So I think that the question here is not to invite more opportunity and less security, but how do we provide that access at a very low cost with the assurances and with the redundancy that we all rely on in our industry.

Mr. ANDRESEN. I think the one positive thing we looked at in lower Manhattan Island is that our conversations with our subscribers about the importance of backup connectivity are much different today than they were in August. And Island has two data centers. They are two hot data centers, so they both link with the primary data center, but it only works for all of our customers, just like Cathy was saying in the New York Stock Exchange's case, if people are actually able to be physically connected. All of our subscribers were connected in New York, and about half were connected to our New Jersey data centers. Now all of the people are connected.

I think the one very good thing, as I said, that has come out of this is that all the brokerage firms, all of the institutions are now painfully aware of what a point of failure truly is; that you need redundant geography, redundant connectivity in terms of carriers, and redundant lines, and you better have redundant hardware as well. It has to be tested, it has to be stressed.

Mr. O'HARA. If I may add, the SEC has changes through its automated review process audit. It has very high standards in ECNs as well. They are contemplated in part—they are ARP'd today, as they say. The SEC, I know, is discussing whether to have full ARP compliance by ECNs as well. What ARP does is make you think about that exact issue, and you have to meet that standard. So from a regulatory standpoint it is covered, and exchanges and ECNs to an extent, and probably the fullest extent of exchanges will have to meet ARP audit standards of the SEC.

Mr. SHADEGG. Anybody else?

Mr. Andresen, you said that we should eliminate any barriers that inhibit fair competition between electronic and traditional markets. Can you expand on that a little bit and elaborate?

Mr. ANDRESEN. Yes sir. Right now, as Ms. Kinney pointed out, ECNs are very—have been very successful in trading a lot of—conducting a lot of business in stocks. We have not been successful doing it in New York. I think that most of that credit has to go to the New York Stock Exchange. They are the preeminent price discovery marketplace in the world.

However, in some ways Island feels that we are still competing with one hand tied behind our back. That is, when someone looks to see who has the best price in IBM, the investor who goes on to their Schwab account or goes onto Yahoo Finance to get their stock quotes doesn't see Island's price included, and that is, in my mind, one of the barriers to competition is that some marketplaces prices are treated as more legitimate than others, when, in fact, the legitimacy comes down to which one is faster, more reliable, more accountable, cheaper and has a greater certainty of execution.

So as I have said earlier, Island actually takes all of our information and gives it out over the Internet for free. Unfortunately, while most professionals see that, and maybe a lot of investors do, when you go and ask where the best price is, you don't see it.

I think it is analogous to a shopping mall. If you go to the shopping mall, you walk in, and you are like me, you like to get out of there fast. You go right to the map at the front. It tells you the men's shoes, B-6, and you go there and you get something to eat and you are out of there. The national market system provides such a map. It tells you where the best price is. Unfortunately, right now Island is treated like, you know, a Burlington Coat Factory across the highway. We hope that people, after they are done shopping, go over there and check us out on their way home. That quasi legitimacy that being in that shopping mall affords, and not having that is a significant competitive disadvantage for us.

Mr. STEARNS. The gentleman's time has expired.

The gentleman from Massachusetts Mr. Markey.

Mr. MARKEY. Thank you, Mr. Chairman very much. And thank you for holding this very, very important hearing.

In a November 14th speech, Peter Vinella, the CEO of PVA International, a Wall Street consulting firm, identified a number of vulnerabilities in the U.S. financial system to a terrorist attack. I would like to walk the panel through three that he highlighted and get your reaction.

Mr. Vinella said, quote, the financial—the U.S. financial system is vulnerable to a number of types of deliberate terrorist attacks. Here are three of the most obvious. No. 1, terrorists could destroy the major telephone switching stations in Midtown and downtown New York. Most of the financial system interacts over a network of dedicated point-to-point phone lines leased by individual firms. Nearly all electronic communications, even dedicated lease lines, use public phone company services and infrastructure. The destruction of a single Verizon switching station near the Trade Center disrupted electronic communications in Manhattan for weeks.

Two, terrorists could place erroneous activity into the financial system. Most electronic trading systems are designed to prevent access by people unauthorized to trade, but nothing prevents a terrorist with the appropriate authorization from sowing enormous confusion in the financial markets. The New York Stock Exchange direct order turnaround trading system, the DOT system, is particularly vulnerable. The New York Stock Exchange members commonly allow their large institutional clients to direct their orders directly to the floor by way of the DOT system. From the New York Stock Exchange point of view, any traffic coming from the member firm is authorized activity. Although member firms are responsible for all activity on the line, they do not monitor their clients' DOT lines and assume that all the traffic on the line is authorized by their clients. A single terrorist working at an institutional client with access to the DOT line could send a high volume of convincingly realistic orders that would trigger a major sell-off. All he would need to do is simply access a trading terminal and a simple password or two.

And three, terrorists could destroy both the primary and disaster recovery sites of major financial institutions. The disaster recovery location of most firms is public knowledge. It wouldn't be difficult to place a truck bomb in New York and in a site in New Jersey.

So I would like to ask our witnesses to deal with this question and begin with you, Ms. Kinney, if we could.

Ms. KINNEY. Since we all can hope to be in business for a long time to come, we will have to address all of these issues. Let me just start with the issue of the telecommunication and switching system. As I said earlier, the Exchange has developed private networks to access and to have a communication with its member firms. So we continue to look at that as a model and perhaps extend that model further, and also to continue to use other sources of connectivity other than simply lease lines or lines—dedicated lines, perhaps even using both virtual private networks as well as the Internet.

So I think to the point, Mr. Markey, that was made here earlier, we all understand the importance of connectivity. We all understand the importance of redundancy, and we all have been—have brought all of that back and fresh and looking at it again to make sure that we can ensure and give confidence to investors that we are going to be there and able to trade, so that all of that that we have done that has worked well we will continue, and that which we found to be an area that could be improved subsequent to September 11 and even to these points is being addressed.

With respect to the activity or unauthorized activity, I would say two things. One is the member firms, when they introduce orders to the New York Stock Exchange, they do have all of those orders coming through their infrastructure today. They are responsible for that order flow. Many of them do actually monitor that. And the Exchange will be putting in some services over time so that the firms can monitor the order flow and the commitment of capital or guarantees that the member firms have extended to those institutions more aggressively. We think that because of this connection, this private connection, between the firms and the Exchange, and the knowledge that the firms have to maintain with who are their customers, and the fact that these orders are going through their infrastructure before they get to our infrastructure, that together from a regulatory perspective we at least will be alert to who our customers are.

That said, I think all of the points that have been raised here certainly are under way and are being addressed by all of us to ensure that we don't have circumstances where unauthorized people are accessing our systems.

And I would say, last but not least, primary and backup services particularly for data sites are being evaluated by all of us. I think you probably are aware and have visited both of our data sites, many of you have. They are about a mile and a half apart as the crow flies. So all of those things are under discussion right now as to whether the Exchange moves to a third or move one of its data sites to a more desirable geographic location.

So that said, all of these things I think very much came out of our experience on September 11. Many of them have been addressed, and many will continue.

Mr. MARKEY. Can I ask you this question, because it is raised by Verizon, which has been making the case that it is really better to have one big monopoly there rather than having competitors there as alternative networks that could be used. What happened to you on that day and subsequent days? Was it helpful to have alternative networks that could be used in addition to Verizon, or would it have been better if just one company, the monopoly as Verizon was saying, is there?

Ms. KINNEY. It didn't matter what might have happened or what might have been better. I think all of us had the experience of relying on Verizon. So we face the challenge of having to work with them to get our customers reestablished in either their new locations or the current locations that were affected.

Mr. MARKEY. But it does matter prospectively. Are you better having one network, or would you prefer to have redundant networks there so that you didn't have to be dependent just on one company, but there would be alternative ways to get business done?

Ms. KINNEY. It seems to me that Verizon has central offices located around the New York City area. I think this is more a question of making sure that everybody has redundancies within, you know, the New York City locations that are not as vulnerable perhaps as what was experienced.

I think, Mr. Markey, also one of the things I said earlier that we all learned was that even if I thought I had Verizon as a primary and MCI as a backup, lots of cables have not been pulled in New York City for a very long time. So we all find ourselves, like it or not, with an issue that has to be addressed. We are reliant on their services.

Mr. MARKEY. I know you are today, but we are also the telecommunications committee, so from my perspective I have always believed that if you had many competitors each providing, you know, service, that if one went down because their location was hit, then could you move the system, basically the economic system of the country, over to others if the New York Stock Exchange is vital in terms of the entire competent functioning. So philosophically would you prefer to have multiple networks or just one in terms of the vulnerability?

Ms. KINNEY. I don't think we have one network. I think you have heard everybody talk about the fact that there are a variety of points, a variety of execution models. We all have——

Mr. MARKEY. So you prefer to have multiple networks.

Ms. KINNEY. We have multiple data sites today. We have two active data sites that support the New York Stock Exchange. I think that where we feel strongly about centralization, leaving the telecommunications side aside for a moment, is that bringing lots of buyers and sellers together in a single point provides benefits for the price discovery model, provides, certainly for us, benefits in terms of information flow. It certainly provides benefits to us in terms of regulation. So we feel very strongly about that.

Back to your other point, I think we will all be searching for ways to insulate ourselves from whatever vulnerabilities or difficulties we experienced on September 11, and we will be looking for providers who do that. Verizon will have to comport with that along with all the other providers of service that exist.

Mr. O'HARA. Congressman Markey, can I answer your question? Yes.

Mr. MARKEY. Yes, you want redundancy. That is all I am looking for.

Let's go down. Yes or no, do you want redundancy; yes or not?

Ms. KINNEY. Yes, of course. We have to.

Mr. MARKEY. First man, could you say yes or no?

Mr. STEINMETZ. Yes.

Mr. ANDRESEN. Yes.

Ms. KINNEY. Yes.

Mr. O'HARA. Yes.

Mr. BANG. Yes.

Mr. RANDICH. Yes.

Mr. JAMAITIS. Absolutely, yes.

Mr. STEARNS. The gentleman's time has expired. We are going to do a second round, so he is welcome to ask his questions——

Mr. MARKEY. It is a yes or no for this question, so it would take no more than 10 seconds, then I could finish up. Do you believe that the institutions that use the DOT currently have appropriate internal controls to address the problems of a terrorist using the DOT to trigger a sell-off?

Mr. STEINMETZ. That actually depends what system they are using. It is not a simple yes or no answer.

Mr. MARKEY. Maybe yes? More yes or no?

Mr. STEINMETZ. I could give a maybe on that.

Mr. ANDRESEN. I will give a Washington answer. I don't have an idea about their system.

Mr. MARKEY. Good.

Ms. KINNEY. It doesn't—terrorists are not——

Mr. STEARNS. Could you talk into the mike?

Ms. KINNEY. The terrorist issue is not the issue. Today every single day everybody shows up and enters orders into the systems and think that the controls are there. Certainly September 11 heightened all of our——

Mr. MARKEY. So you think the controls are there. So you don't believe that institutions are vulnerable to this kind of attack?

Ms. KINNEY. I think that institutions will need to be more alert, but——

Mr. MARKEY. You don't think they are vulnerable.

Ms. KINNEY. We face these issues every single day.

Mr. BANG. We employ risk management systems, so for the most part we are protected against this.

Mr. MARKEY. You are protected.

Mr. O'HARA. Yes.

Then I would confirm the same thing with us. Again, I will point back before you came in Congressman, we talked about the ARP review that is an audit done by the SEC. This is picked up by it; that if they need to tweak that, maybe that is a place to start.

Mr. MARKEY. Are more safeguards needed?

Mr. O'HARA. I defer to the SEC. They do come in and audit these types of things.

Mr. MARKEY. The SEC did not believe there was a problem with program trading in the summer of 1987. They all testified here, every one of the exchanges in the SEC, we have safeguards in place. So this is going to be the hearing, by the way, if anything happens in the future, you are all saying, no problem. So just so you know, they all said no problem, which got them in a lot of problems 3 months later. No problems, they all said. So you are on record now for this issue the way they were for program trading.

Mr. Randich.

Mr. RANDICH. NASDAQ uses real-time surveillance systems for this purpose and others, but we need to be continuously cautious.

Mr. MARKEY. Mr. Jamaitis.

Mr. JAMAITIS. Any safeguards would be very system-dependent, so it is not a simple yes or no. As technology improved our ability to detect these type of things——

Mr. MARKEY. Does every institution have the internal controls necessary to make sure there can't be an attack? Do you agree that they do or don't?

Mr. JAMAITIS. I agree there is probably room for improvement.

Mr. MARKEY. Do you know institutions that do not have controls that are sufficient to protect against it?

Mr. JAMAITIS. No.

Mr. MARKEY. That is all good news to us that terrorists cannot really get through the system.

Mr. STEARNS. We will have a second round. We would like to have your questions. Let me open up. I will just take a short amount of time.

Ms. Kinney had mentioned that the ECNs are only about 3 percent of the New York Stock Exchange. And let me ask Mr. Randich, what percent is ECN of NASDAQ?

Mr. RANDICH. Thirty-five to 40 percent.

Mr. STEARNS. So they are 40 percent. Now, why is it 40 percent with the NASDAQ, but only 3 percent of the New York Stock Exchange?

Mr. RANDICH. I can speak for NASDAQ as NASDAQ has a very open democrat architecture. We allow free access to many participants, and the ECNs have thrived in that environment.

Mr. STEARNS. So most of the ECNs were started in the late 1990's, 1996. So in that short amount of time, they have gotten 40 percent of the business of NASDAQ, but they are only stuck at 3 percent at the New York Stock Exchange. So, Mr. O'Hara, what is your comment why?

Mr. O'HARA. I think if you refer back to our testimony and others at a data hearing, it is fairly—there is a primary answer, and then there are some side answers. But the primary answer is that there are legal and regulatory hurdles, some large, some have come down a little bit, for ECNs to compete in listed——

Mr. STEARNS. With the New York Stock Exchange, but they are not with NASDAQ.

Mr. O'HARA. That is correct.

Mr. STEARNS. So all these regulatory hurdles are difficult for you to do business on the New York Stock Exchange.

Mr. O'HARA. We have found it that way. Now we are graduating to Exchange status ourselves and would certainly like to change the clubby atmosphere that we call a fraternity house. There are certain blackballs, who is allowed in and who gets paddled for doing bad things.

Mr. STEARNS. So you are saying that the New York Stock Exchange is a fraternity with blackballs?

Mr. O'HARA. No. I am saying that the national market assists the committees, the ITS committee and CTA, and we have had our battles there. If one—I believe there is eight or nine people on that.

If one of them doesn't like the way you look that day, they pull out blackball, just the fraternity house, and you are done.

Mr. STEARNS. Would you put into our record a letter outlining what you think the regulatory hurdles are?

Mr. O'HARA. I certainly would. Thank you.

Mr. STEARNS. Ms. Kinney, obviously, you would like to respond to that?

Ms. KINNEY. I don't know about fraternities, but——

Mr. STEARNS. I think being a fraternity might not be the right word.

Ms. KINNEY. I think the New York Stock Exchange provides a platform for competition. I keep referring to the chart to the left that was brought by the same gentleman who said that the barriers are high. I would say that all of these market models are free to compete with the New York Stock Exchange every day.

Mr. STEARNS. Last year, the Subcommittee Chairman of Finance and Hazardous Materials wrote to the SEC on an issue that seemed to affect the ECNs. I mentioned that in my opening statement. What I would like to do is have someone explain to me what the Consolidated Tape Association, or CTA, is. And just maybe the New York Stock Exchange could start; and, Mr. Steinmetz, you could talk in terms of if you are having access to the information that is generated by the stock trades and have the counterpoint to Ms. Kinney—is that possible—that you could explain what the Consolidated Tape Association, or the CTA, is?

Ms. KINNEY. I will do my best.

Coming out of the 1975 act's amendment, the first priority was to provide information to the marketplace transparency about the markets and the bids and offers and trades that were taking place in those variety of marketplaces or market centers. So the Consolidated Tape Association was formed as the first block.

Mr. STEARNS. What year was that?

Ms. KINNEY. That was in the 1970's. Following—1979, just following the 1975 act's amendments. It was the first piece, the intermarket trading system coming just after that.

But, again, it was to provide the marketplaces with an opportunity to centralize their information to provide a summarized best bid and offer and to provide the transaction information that was occurring in the various market centers.

You know, the CTA has operated since that time. Hearings have been held—in fact, Mr. Seligman, at the direction of the SEC, held quite a number of—I won't say they were called hearings but meetings about how to make information, what about transparency, what about the Consolidated Tape Association, how to make that information—and those recommendations certainly have been put into the marketplace over the last several months.

Mr. STEARNS. Mr. Steinmetz, do you think there should be changes to the CTA? And, if so, what should they be?

Mr. STEINMETZ. Well, yes, there probably should be some changes to the CTA. I think the idea is that market data has similar issues to the ability to execute trades, and that is the ability to do it in different places or—as far as executing trades and as far as delivering the data, it should be able to be delivered dependent upon the network that actually has the data. So, for instance, if Instinet has certain data order flow quotes and orders that could be displayed, there should be some participation from the participant who actually has that data and order flow to be able to get something out of it.

It, like the standard systems in general, should again go down the whole idea of the network effect rather than the single point so that there can be access to it and reward from it from multiple participants and not just one central location.

Mr. STEARNS. Mr. Andresen, do you have any comments on what, if any, regulatory change should be made to the CTA?

Mr. ANDRESEN. Yes, sir. One thing I want to point out, it is important to draw distinction between the New York Stock Exchange and the trading of the New York Stock Exchange listed stocks. The New York Stock Exchange in itself is not keeping us down on Broad Street.

What is going on is the New York Stock Exchange lists these securities. They trade primarily on the New York. New York, as was outlined here today, does the majority of the trading of the stock. However, trading is facilitated in other places like Island and Archipelago and Instinet.

So to continue with my analogy about a shopping mall, it would just be like having—the New York Stock Exchange is just like Macy's, the enormous store at the end of the shopping mall, and there are lots of little stores around it. The consolidated tape is that shopping mall, takes the prices from the different market places and right now primarily from the New York Stock Exchange.

The issue with the consolidated tape in our mind at Island is that, for us to gain admission into the consolidated tape and now have our prices in the shopping mall, we then must also be a part of the intermarket trading system. You know, this, as I have outlined earlier, would be akin to if someone came in to buy some orange juice in our store in the shopping mall and someone had a better advertised price in another part of the shopping mall, it would be my obligation to grab them and put them in one of the carts and wheel them down to the other store and say I can't sell this to you, you have to go somewhere else.

Nowhere else in our economy—if I go into 7-Eleven to buy a quart of milk, I do it at 2 in the morning, I don't want 7-Eleven to send me off to Sam's Club in the boonies which isn't open for 8 hours, which I have to pay $50 to get in and where the milk is in five gallon jugs.

So different customers are going to want different services. I think we have—there is unanimity——

Mr. STEARNS. So the interim market trading system is what really should be addressed, rather than the Consolidated Tape Association, in your opinion.

Mr. ANDRESEN. That is correct.

Mr. STEARNS. And in that area is where we need the reform.

Mr. ANDRESEN. That is correct.

Mr. STEARNS. Does anyone disagree or agree with that?

Mr. O'HARA. I think a couple things, a couple historical markers here. In 1975 Congress, from a law standpoint, called on the SEC to build these structures. I think it was well intended and worked well until the days when we are now competing. These are competitors to primary markets.

The SEC—and I think this is a good education for all—they held a yearlong committee headed by Dean Joel Seligman and discussed this issue where all the market centers, including probably everyone at this table, submitted white papers on this very issue.

I think—from our standpoint, I think the New York Stock Exchange actually agreed with this, that the whole idea of consolidating reflecting sale of market data should be deregulated. And today it is heavily regulated. It is—there is a quasi-monopolistic pricing, if you will.

If there is something to be looked at here it is the deregulation of this conglomeration of national market committees, i.e., ITS, CTA, and especially, you know, how they govern, again going back to this one black ball governance system where, if someone doesn't like you that day, you are done.

Mr. STEARNS. I think my time is up. And now the distinguished ranking member from New York, who will have his questions.

Mr. TOWNS. Thank you very much, Mr. Chairman.

I think earlier—I think it might have been you, Mr. Bang, that mentioned, in terms of the public policy discussion in terms of decentralization versus centralization, I noticed that you sort of came down on the side of decentralization. However, I would think that the trend would be the centralization with a back-up system, wouldn't it, in terms of efficiency?

Mr. BANG. No. I would argue that decentralization is a better redundancy and provides for more competition and innovation in the long term because the centralization implies one counter party—one central point of failure and one service provider, whereas a decentralized marketplace implies multiple liquidity hubs operated by competing market centers offering the end investor an alternative destination for trading the same securities. So I would say decentralization is the preferable way to go.

Mr. TOWNS. Ms. Kinney, can I ask you to respond to that as well?

Let me just say, first of all, before I say that, congratulations on your being the future president of the Exchange. It is a great milestone and can only happen in New York. And also to give you an opportunity to speak a word on behalf of the sororities.

Ms. KINNEY. I think that one of the things that you are seeing in the marketplace today is a lot of consolidation among the various participants. I think Mr. O'Hara talked about the consolidation of Archipelago and REDIBook, and I think all of us are clearly looking at how to provide efficiency, a very cost-effective access to our respective models but to do that in a way that ensures that there is both redundancy as well as certainty to the customers who choose us. So I think we have to have—to Mr. Markey's question—redundancy and decentralization of our data sites and a variety of other things.

That said, I think you are correct in the sense that we feel that the more centralization for order flow and a variety of other things the better investors and customers are served.

Mr. BANG. Decentralization of order flow is essentially achieved today through a virtual network, and the key is to get these liquidity centers representing their clients' interests in quote montage. If you have that transparency and it is disseminated and the connectivity is provided, then essentially you have a virtual centralization of liquidity, which is better than a physical centralization of that liquidity.

Mr. TOWNS. Thank very much, Mr. Chairman.

On that note, I yield back; and I wanted to congratulate you on having this hearing. I think it is important that we are able to get this kind of information. And I wanted to let you know the SEC is in the room.

Mr. STEARNS. I thank my distinguished colleague.

Let me conclude the hearing by again complimenting the exchanges for getting back into business so quickly and the ECNs for their rapid deployment during this September 11 crisis. As a result of this hearing, there is a lot of perhaps policy issues that the committee should look at.

I thank you for your participation and hope all of you have a happy holiday.

[Whereupon, at 12:30 p.m., the subcommittee was adjourned.]

○

APEX
TP
9780160669156
76901945 – 59